NUMERICAL ANALYSIS

An Introduction

NUMERICAL ANALYSIS

An Introduction

Lars Eldén
Linde Wittmeyer-Koch

Department of Mathematics
Linköping University
Linköping, Sweden

ACADEMIC PRESS, INC.

Harcourt Brace Jovanovich, Publishers

Boston San Diego New York
London Sydney Tokyo Toronto

Copyright © 1990 by Academic Press, Inc.
All rights reserved.
No part of this publication may be reproduced or
transmitted in any form or by any means, electronic
or mechanical, including photocopy, recording, or
any information storage and retrieval system, without
permission in writing from the publisher.

ACADEMIC PRESS, INC.
1250 Sixth Avenue, San Diego, CA 92101

United Kingdom Edition published by
ACADEMIC PRESS LIMITED
24-28 Oval Road, London NW1 7DX

Library of Congress Cataloging-in-Publication Data
Eldén, Lars, date
    [Numerisk analys. English]
    Numerical analysis : an introduction / Lars Eldén, Linde Wittmeyer
-Koch.
        p. cm.
    Rev. translation of: Numerisk analys.
    Includes bibliographical references.
    ISBN 0-12-236430-9 (alk. paper)
    1. Numerical analysis. I. Wittmeyer-Koch, Linde, 1939–
II. Title.
QA297.E4213 1990   89-71379
519.4—dc20   CIP

Printed in the United States of America

90 91 92 93   9 8 7 6 5 4 3 2 1

Contents

Preface

This book is intended as a textbook for an introductory course in numerical analysis. The level is suitable for the advanced undergraduate. We believe that the book is accessible to a rather wide audience, since most of the mathematics prerequisite should be covered in first-year calculus and algebra courses.

The aim of this book is to give an introduction to some of the basic ideas in numerical analysis. We have chosen to cover a comparatively wide range of material, including classical algorithms for the solution of nonlinear equations and for linear systems, methods for interpolation, integration and approximation. We have also tried to give an introduction to some areas, which we think are important in the applications that a science or engineering student will meet. These areas include computer arithmetic (floating point) and standard functions, splines and finite elements. In a few areas (approximation and linear systems of equations) we have chosen to give a somewhat more comprehensive presentation.

In our opinion a broad introductory course should emphasize the understanding of methods and algorithms. In order to really grasp what is involved in numerical computations, the student must first perform computations using a simple calculator. We supply a number of exercises intended for this. To further help the student in the learning process we recommend the use of high level programming systems (like Matlab or Mathematica) for more realistic computer assignments than we have given in this book.

This book is a revision of a corresponding book earlier published in Swedish. We are grateful to many of our colleagues at the Department of Mathematics, Linköping University, for suggesting several improvements, and for correcting errors. In particular we would like to mention Tommy Elfving, Jan Eriksson, and Ulla Ouchterlony. Ingegerd Skoglund has constructed most of the TEX macros we have used.

Linköping, January 1990

Lars Eldén
Linde Wittmeyer–Koch

1   Introduction

1.1   Mathematical Models and Numerical Approximations

Mathematical models are basic tools in scientific problem solving. Typically, some fundamental natural laws are used to derive one or several equations, which model the process that is being studied. Through the mathematical treatment of the equations, answers can be found to questions that are posed in connection with the problem area. This is illustrated schematically in Figure 1.1.1.

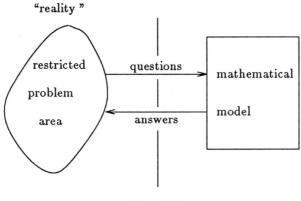

Figure 1.1.1

It is important to remember that in many cases the problem area described by a mathematical model is very narrow. Further, there are often simplifications in the assumptions. Therefore, the mathematical model is not an exact description of reality, and the answers that it produces must be checked and compared with experimental results.

We shall illustrate the concept of a mathematical model using an example from structural mechanics. It is not our aim to give a description that is comprehensive from the point of view of mechanics. Nor will the mathematical aspects be treated in detail. Rather, we shall use a relatively simple example to show that mathematical analysis is not enough to answer the questions that are relevant from the viewpoint of an application.

Consider a beam of length 1, rigidly built-in at both ends.

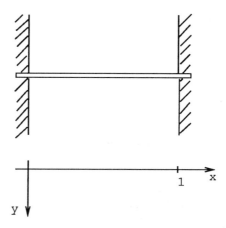

Figure 1.1.2

The following question is interesting: How much will the beam deflect under a certain load?

It is also important to know bending moments, shear forces, etc. Information about these can be found as byproducts when the question about the deflection is answered.

We introduce a coordinate system according to the figure, and let $y(x)$ denote the deflection of the beam under a certain load. If the load is $q(x)$, not necessarily constant over the length of the beam, then (under certain assumptions) $y(x)$ satisfies the differential equation

$$\frac{d^2}{dx^2}\left(EI(x)\frac{d^2y}{dx^2}\right) = q(x), \qquad 0 < x < 1, \qquad (1.1.3a)$$

with boundary conditions

$$y(0) = y'(0) = y(1) = y'(1) = 0; \qquad (1.1.3b)$$

E is a material constant (Young's modulus of elasticity) and $I(x)$ is the moment of inertia of the cross-section of the beam. Note how the boundary conditions describe the fact that the beam is rigidly built-in at both ends.

(1.1.3a) is called the beam equation and (1.1.3) is an example of a **boundary value problem** for a fourth-order differential equation. If we can solve (1.1.3), i.e., determine the function $y(x)$ that satisfies (1.1.3), then we have answered the question about the deflection of the beam.

In some cases, it is easy to solve the boundary value problem analytically. Assume, e.g., that the coefficients and the load are constant

$$I(x) = I_0, \quad q(x) = q_0,$$

and put

$$Q = \frac{q_0}{EI_0}.$$

We immediately see that

$$y(x) = \frac{Q}{24}x^4 + Ax^3 + Bx^2 + Cx + D,$$

satisfies the differential equation (1.1.3a) for arbitrary A, B, C and D. From the boundary conditions we can determine A, B, C and D and get

$$y(x) = \frac{Q}{24}(x^4 - 2x^3 + x^2) = \frac{q_0}{24EI_0}(x^4 - 2x^3 + x^2). \qquad (1.1.4)$$

With the analytical solution (1.1.4), we can determine the deflection of the beam for any value of x and for any value of the load q_0.

The problem is slightly more complicated if the load $q(x)$ is not constant, but it can still be solved analytically, if we can determine successive primitive functions of primitive functions of q.

The problem becomes considerably more complicated if we assume that the beam rests on an elastic foundation.

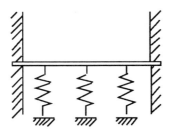

Figure 1.1.5

The deflection of the beam now satisfies the equation

$$\frac{d^2}{dx^2}\Big(EI(x)\,\frac{d^2y}{dx^2}\Big) + k(x)y = q(x), \qquad (1.1.6)$$

with boundary values (1.1.3b). $k(x)$ is the foundation spring function and is assumed to be a continuous function of x.

Using mathematical analysis, one can prove theorems about the existence and uniqueness of solutions of (1.1.6), but in general no explicit solution can be found. An explicit solution is a formula, where for each set of values of the relevant parameters and for each x we can easily compute the corresponding function value $y(x)$.

Thus, we have a mathematical model which is a good description of the problem under study, but it does not give us a direct answer to our question. Now there are essentially two alternatives: either we can make simplifications in the model (e.g., assume that the coefficients are constant) so that an analytical solution can be determined, or we can introduce **numerical approximations** in the equation. In many cases, the first alternative is inadequate, as the model may become so bad that the answers are no longer reliable. Errors are introduced in the second alternative also, but here the answers are more reliable since it is possible to estimate how the approximations affect the accuracy of the solution. The two alternatives are illustrated in Figure 1.1.7.

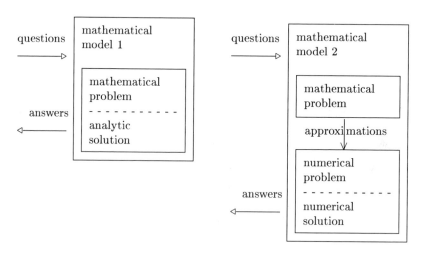

Figure 1.1.7

We will now sketch how numerical approximations can be made in the differential equation (1.1.6). For a detailed description, see Chapter 10.

For many purposes it is sufficient to compute approximations of the solution at a finite number of points $(x_i)_{i=1}^n$ in the interval $[0, 1]$. We approximate the derivatives in the differential equation (1.1.6) and the boundary conditions (1.1.3b) by difference quotients, e.g.,

$$y''(x_i) \approx \frac{y(x_{i+1}) - 2y(x_i) + y(x_{i-1})}{h^2};$$

here we have assumed that the partitioning of the interval is equidistant, i.e., $x_{i+1} - x_i = h$. In this way, the differential equation (1.1.6) is replaced by a system of linear equations

$$Ay = q; \tag{1.1.8}$$

the components of the vector y are approximations of the function values

$$y_i \approx y(x_i), \qquad i = 1, 2, \ldots, n;$$

the elements of the matrix A depend on the coefficients in (1.1.6) (the material properties), and the right hand side q is a vector of values of the function $q(x)$ (the load).

The linear system of equations (1.1.8) is called a **discretization** of the differential equation (1.1.6) with boundary conditions (1.1.3b). The system of equations can be solved using Gaussian elimination.

Thus, we have replaced the mathematical problem of solving a differential equation by the numerical problem of solving a linear system of equations. We will now make the notion of a numerical problem somewhat more precise:

A **numerical problem** is a clear and unambiguous description of the functional connection between **input data**, i.e., the "independent variables" of the problem, and **output data**, i.e., the desired results. Input data and output data consist of a *finite number* of real quantities.

It is obvious that the problem of solving (1.1.8) can be considered as a numerical problem. Input data are the coefficients of the matrix A and the right hand side q; output data are the components of the vector y.

We stated that (1.1.8) can be solved using Gaussian elimination. That is an example of an algorithm.

An **algorithm** for a numerical problem is a complete description of a finite number of well-defined operations, through which each permissible input data vector is transformed into an output data vector.

By operations, here we mean arithmetic and logical operations and previously defined algorithms. An algorithm can be described loosely or in
great detail. A comprehensive description is obtained when an algorithm
is formulated using a programming language.

The objective of **numerical analysis** is to construct and analyze numerical methods and algorithms for the solution of problems in science and
technology. In connection with the above discussion, we give here a few
examples of interesting questions in numerical analysis:

- How large is the discretization error when the boundary value problem
  is approximated by (1.1.8), i.e., how large are the errors $y_i - y(x_i)$?
- How long does it take to solve (1.1.8) using a certain computer?
- How do the rounding errors of the computer arithmetic influence the
  accuracy of the solution?

Exercises

1. Derive (1.1.4) and sketch the curve for some value of Q. What is the
   maximal deflection of the beam?

2. Find the chapter on beams in a textbook in structural mechanics and
   write the beam equation using the notation of that book.

3. What are the boundary conditions for (1.1.3a) if the beam is freely sup-
   ported at both ends. Determine an analytical solution for this case
   (assume constant coefficients).

References

The definitions of a numerical problem and algorithm are taken from

G. Dahlquist and Å. Björck, *Numerical Methods*, Prentice–Hall, Englewood Cliffs, New Jersey, 1974.

This textbook is recommended for a more extensive course in numerical
analysis.

2   Error Analysis and Computer Arithmetic

In Chapter 1, we illustrated how approximations are introduced in the solution of mathematical problems that cannot be solved exactly. One of the most important tasks in **numerical analysis** is to estimate the accuracy of the result of a numerical computation. In this chapter, we discuss different sources of error that affect the computed result and we derive methods for error estimation. In particular, we describe some properties of computer arithmetic. A standard for floating point arithmetic was adopted by IEEE in 1985. We give a brief presentation of the standard and its implementation in a floating point processor.

2.1   Sources of Error

There are essentially three types of errors that affect the result of a numerical computation:

1. **Errors in the input data** are often inevitable. The input data can be the result of measurements with a limited accuracy, or real numbers, which must be represented with a fixed number of digits.
2. **Rounding errors** arise when computations are performed using a fixed number of digits in the operands.
3. **Truncation errors** arise when "an infinite process is replaced by a finite one", e.g., when an infinite series is approximated by a partial sum, or a function is approximated by a polynomial.

Truncation errors will be discussed in connection with the different numerical methods. In this chapter, we study the other two sources of error.

The different types of errors are illustrated in Figure 2.1.1, which relates to the discussion in Chapter 1.

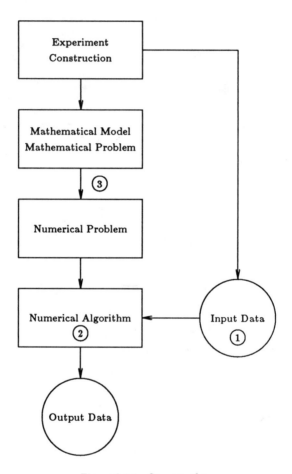

Figure 2.1.1   Sources of error.

The following notation will be used:

R_X	error in the result coming from input data errors,
R_{XF}	error in the result coming from errors in the function values used,
R_B	rounding error,
R_T	truncation error.

R_{XF} is a special case of R_X.

2.2   Basic Concepts

Let a denote an **exact value**, and $\bar{a}$ an **approximation** of a. E.g.,

$$a = \sqrt{2}, \quad \bar{a} = 1.414.$$

We introduce the following definitions.

> **Absolute error** in $\bar{a}$: $\Delta a = \bar{a} - a,$
>
> **Relative error** in $\bar{a}$: $\dfrac{\Delta a}{a}, \quad (a \neq 0).$

In the above example, we have

$$\Delta a = 1.414 - \sqrt{2} = -0.0002135\ldots,$$

$$\frac{\Delta a}{a} = \frac{-0.0002135\ldots}{\sqrt{2}}.$$

In many cases, we only know a bound of the magnitude of the absolute error of an approximation. Also, it is often practical to give an estimate of the magnitude of the absolute and relative error, even if more information is available. E.g.,

$$|\Delta a| \leq 0.00022,$$

$$\left|\frac{\Delta a}{a}\right| \leq 0.00016,$$

or

$$|\Delta a| \leq 0.0003,$$

$$\left|\frac{\Delta a}{a}\right| \leq 0.0002$$

(note that we must always round upwards in order that the inequalities shall hold). Relative errors are often given in percentages. In the above example, the error is 0.02% at most.

The following three statements are equivalent.

$$\bar{a} = 1.414, \quad |\Delta a| \leq 0.22 \cdot 10^{-3},$$

$$a = 1.414 \pm 0.22 \cdot 10^{-3},$$

$$1.41378 \leq a \leq 1.41422.$$

There are two ways to reduce the number of digits in a numerical value: **rounding** and **chopping**. We first consider rounding of decimal numbers to t decimals.

The tth decimal is raised by 1 if the part of the number in positions to the right of the tth decimal is greater than $0.5 \cdot 10^{-t}$. If it is less than $0.5 \cdot 10^{-t}$, then the tth decimal is not changed. In the limit case, when the number to the right of the tth decimal is equal to $0.5 \cdot 10^{-t}$ exactly, then the tth decimal is raised by 1 if it is odd, but is left unchanged if it is even. This is called **rounding to even** in the limit case. Chopping, on the other hand, means that all decimals to the right of the tth decimal are simply eliminated.

Example 2.2.1   Rounding to 3 decimals:

1.23767	is rounded to	1.238,
0.774432	"	0.774,
6.3225	"	6.322,
6.3235	"	6.324.

Chopping to 3 decimals:

0.69999	is chopped to	0.699.

It is important to remember that errors are introduced when numbers are rounded or chopped. From the above rules, we see that when a number is rounded to t decimals, then the error is $0.5 \cdot 10^{-t}$ at most, while the chopping error can be 10^{-t}. Note that chopping errors are systematic: the magnitude of the result is always smaller that that of the original number. When an approximate value is rounded or chopped, then that error must be added to the error bound.

Example 2.2.2   Let $b = 11.2376 \pm 0.1$. Here, it is not meaningful to give four decimals, since the absolute error can be one unit in the first decimal. Therefore, we round to one decimal: $b_{\text{rounded}} = 11.2$. The rounding error is

$$|R_B| = |b_{\text{rounded}} - \bar{b}| = |11.2 - 11.2376| = 0.0376 < 0.04.$$

We must add the bound for the rounding error to the original error bound:

$$b = 11.2 \pm 0.14.$$

This is easily seen if we write the original approximate value and error bound as an interval:

$$11.1376 \le b \le 11.3376.$$

When taking 11.2 as a new approximate value, we must compensate for this:
$$11.06 \leq b \leq 11.34$$
(observe that $11.1 \leq b \leq 11.3$ is not necessarily true).

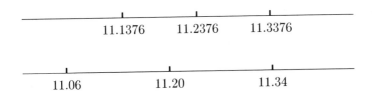

Rounding and chopping in other number systems than the decimal system are completely analogous.

The following definitions are closely related to the concepts absolute and relative error.

If $|\Delta a| \leq 0.5 \cdot 10^{-t}$ then the approximate value $\bar{a}$ is said to have t **correct decimals**.

In an approximate value with t correct decimals, all digits in positions with unit larger than or equal to 10^{-t} are called **significant digits** (leading zeros are not counted; they only indicate the position of the decimal point).

The definitions can easily be modified so that they cover the case when the magnitude of the absolute error is larger than 0.5.

Example 2.2.3   From the definitions, we have

Approximation with Error Bound	Correct Decimals	Significant Digits
$0.001234 \pm 0.5 \cdot 10^{-5}$	5	3
$56.789 \pm 0.5 \cdot 10^{-3}$	3	5
210000 ± 5000		2

Note that the approximation

$$a = 1.789 \pm 0.005$$

has two correct decimals in spite of the fact that the exact value may be 1.794.

2.3   Error Propagation

When approximate values are used in computations then, of course, their errors will give rise to errors in the results. We shall derive some simple methods for estimating how errors in the data are propagated in computations.

In practical applications, error analysis is often closely related to the technology for constructing a device and measuring physical quantities.

Example 2.3.1   The efficiency of a certain type of solar collector can be computed from the formula

$$\eta = K \frac{\dot{V} T_{\text{diff}}}{I},$$

where K is a constant known to high accuracy, $\dot{V}$ denotes volume flow, T_{diff} denotes temperature difference between ingoing and outgoing fluid, and I denotes irradiance. How accurately we can measure $\dot{V}$, T_{diff} and I depends on the technology used in the solar collector. Assume that we have computed the efficiency for two solar collectors, S1 and S2, and that we can estimate the errors of the data as shown in the table.

Collector	S1	S2
$\dot{V}$	1.5 %	0.5 %
T_{diff}	1 %	1 %
I	3.6 %	2 %

If the approximate efficiencies are 0.76 and 0.70 for S1 and S2, respectively, can we then be sure that S1 is more efficient than S2?

We shall return to this example later on, when we have derived mathematical tools that help us answer the question.

First, assume that we shall compute $f(x)$, where f is a differentiable function; further assume that we know an approximation of x with an error bound: $x = \overline{x} \pm \epsilon$. If f is monotone (increasing or decreasing), then we can estimate the propagated error simply by computing $f(\overline{x} + \epsilon)$ and $f(\overline{x} - \epsilon)$:

$$|\Delta f| = |f(\overline{x}) - f(x)| \leq \max(|f(\overline{x} + \epsilon) - f(\overline{x})|, |f(\overline{x} - \epsilon) - f(\overline{x})|).$$

A more generally applicable method is to use the **mean value theorem**.

Theorem 2.3.2 If the function f is differentiable, then there is a point ξ between $\overline{x}$ and x such that

$$\Delta f = f(\overline{x}) - f(x) = f'(\xi)(\overline{x} - x).$$

When we use the mean value theorem for practical error estimation, we usually evaluate the derivative for $x = \overline{x}$ and add something to the error bound to be on the safe side.

Example 2.3.3 We shall compute $f(a) = \sqrt{a}$, for $a = 2.05 \pm 0.01$. The mean value theorem gives

$$\Delta f = f'(\xi)\,\Delta a = \frac{1}{2\sqrt{\xi}}\,\Delta a.$$

We can estimate

$$|\Delta f| \lesssim \frac{1}{2 \cdot \sqrt{2.05}}\,|\Delta a| \leq \frac{0.01}{2\sqrt{2.05}} \leq 0.0036.$$

($\lesssim$ shall be read "less than or approximately equal to".) In general, there are more than one datum in a computation, all approximations. We first examine error propagation for addition, subtraction, multiplication and division.

Let $y = x_1 + x_2$, and assume that we know approximations $\overline{x}_1$ and $\overline{x}_2$. From the definition of absolute error, we get

$$\Delta y = \overline{y} - y = \overline{x}_1 + \overline{x}_2 - x_1 - x_2 = \Delta x_1 + \Delta x_2.$$

If we only know bounds for the absolute errors in x_1 and x_2, we must take absolute values and use the triangle inequality

$$|\Delta y| = |\Delta x_1 + \Delta x_2| \le |\Delta x_1| + |\Delta x_2|.$$

When we do the same for the subtraction $y = x_1 - x_2$, we get

$$\Delta y = \Delta x_1 - \Delta x_2,$$

and the estimate

$$|\Delta y| \le |\Delta x_1| + |\Delta x_2|.$$

We summarize the results for addition and subtraction.

	$y = x_1 + x_2,$	$y = x_1 - x_2,$
Absolute error:	$\Delta y = \Delta x_1 + \Delta x_2,$	$\Delta y = \Delta x_1 - \Delta x_2,$
Error bound:	$|\Delta y| \le |\Delta x_1| + |\Delta x_2|,$	$|\Delta y| \le |\Delta x_1| + |\Delta x_2|.$

This can easily be generalized to an arbitrary number of data. For $y = \sum_{i=1}^{n} x_i$, we get

$$|\Delta y| \le \sum_{i=1}^{n} |\Delta x_i|.$$

Next consider multiplication $y = x_1 x_2$. We get

$$\Delta y = \bar{x}_1 \bar{x}_2 - x_1 x_2 = (x_1 + \Delta x_1)(x_2 + \Delta x_2) - x_1 x_2$$
$$= x_1 \Delta x_2 + x_2 \Delta x_1 + \Delta x_1 \Delta x_2.$$

If the relative errors in $\bar{x}_1$ and $\bar{x}_2$ are small, then we can disregard the last term. Now it is more convenient to consider relative errors:

$$\frac{\Delta y}{y} \approx \frac{\Delta x_1}{x_1} + \frac{\Delta x_2}{x_2};$$

and if we take absolute values and use the triangle inequality,

$$\left| \frac{\Delta y}{y} \right| \lesssim \left| \frac{\Delta x_1}{x_1} \right| + \left| \frac{\Delta x_2}{x_2} \right|.$$

By an analogous argument for division $y = x_1/x_2$, we get

$$\frac{\Delta y}{y} \approx \frac{\Delta x_1}{x_1} - \frac{\Delta x_2}{x_2},$$

$$\left|\frac{\Delta y}{y}\right| \lesssim \left|\frac{\Delta x_1}{x_1}\right| + \left|\frac{\Delta x_2}{x_2}\right|.$$

We summarize:

	$y = x_1 \cdot x_2,$	$y = x_1/x_2,$
Relative error:	$\frac{\Delta y}{y} \approx \frac{\Delta x_1}{x_1} + \frac{\Delta x_2}{x_2},$	$\frac{\Delta y}{y} \approx \frac{\Delta x_1}{x_1} - \frac{\Delta x_2}{x_2},$
Error bound:	$\left\|\frac{\Delta y}{y}\right\| \lesssim \left\|\frac{\Delta x_1}{x_1}\right\| + \left\|\frac{\Delta x_2}{x_2}\right\|,$	$\left\|\frac{\Delta y}{y}\right\| \lesssim \left\|\frac{\Delta x_1}{x_1}\right\| + \left\|\frac{\Delta x_2}{x_2}\right\|.$

Now we can solve the solar collector problem.

Example 2.3.4 The error propagation formulas for multiplication and division give

$$\left|\frac{\Delta \eta}{\eta}\right| \leq \left|\frac{\Delta \dot{V}}{\dot{V}}\right| + \left|\frac{\Delta T_{\text{diff}}}{T_{\text{diff}}}\right| + \left|\frac{\Delta I}{I}\right|.$$

For S1, we get

$$\left|\frac{\Delta \eta}{\eta}\right| \leq (1.5 + 1 + 3.6) \cdot 10^{-2} = 6.1 \cdot 10^{-2},$$

which leads to

$$|\Delta \eta| \leq 0.76 \cdot 6.1 \cdot 10^{-2} < 0.047.$$

Thus, the efficiency for S1 is in the interval

$$0.713 \leq \eta \leq 0.807.$$

The corresponding computation for S2 gives

$$0.675 \leq \eta \leq 0.725.$$

Since the intervals overlap, we cannot be sure that S1 is better.

When investigating error propagation for the evaluation of a function f of n variables $x_1, x_2, \ldots, x_n$, one can use the following generalization of the mean value theorem.

Theorem 2.3.5   If the real valued function f is differentiable in a neighbourhood of the point $x = (x_1, x_2, \ldots, x_n)$, and $x + \Delta x$ is a point in that neighbourhood, then there is a number θ, $0 < \theta < 1$, such that

$$\Delta f = f(x + \Delta x) - f(x) = \sum_{k=1}^{n} \frac{\partial f}{\partial x_k}(x + \theta \, \Delta x) \, \Delta x_k.$$

Proof.   Define the function $F(t) = f(x + t \, \Delta x)$. The mean value theorem for a function of one variable and the chain rule give

$$\Delta f = F(1) - F(0) = F'(\theta) = \sum_{k=1}^{n} \frac{\partial f}{\partial x_k}(x + \theta \, \Delta x) \, \Delta x_k.$$

∎

When using this theorem in practice, one usually evaluates the partial derivatives for $x = \bar{x}$ (the approximation). When there are only bounds for the errors in the argument x, one can get a bound for Δf by using the triangle inequality.

General error propagation formula:

$$\Delta f \approx \sum_{k=1}^{n} \frac{\partial f}{\partial x_k} \Delta x_k.$$

Maximal error bound:

$$|\Delta f| \lesssim \sum_{k=1}^{n} \left| \frac{\partial f}{\partial x_k} \Delta x_k \right|.$$

Example 2.3.6   Let $y = \sin(x_1^2 x_2)$, where $x_1 = 0.75 \pm 10^{-2}$ and

$x_2 = 0.413 \pm 3 \cdot 10^{-3}$. The maximal error bound gives

$$|\Delta y| \lesssim \left|\cos(x_1^2 x_2) \cdot 2x_1 x_2 \, \Delta x_1\right| + \left|\cos(x_1^2 x_2) \, x_1^2 \, \Delta x_2\right|$$
$$\lesssim 0.974 \cdot 2 \cdot 0.75 \cdot 0.413 \cdot 10^{-2} + 0.974 \cdot 0.5652 \cdot 3 \cdot 10^{-3}$$
$$\leq 0.77 \cdot 10^{-2}.$$

The approximate value of y is

$$\overline{y} = \sin(0.75^2 \cdot 0.413) = 0.230229 \pm 0.5 \cdot 10^{-6}.$$

If we round this to 0.23, we get a rounding error less than $0.03 \cdot 10^{-2}$. Thus, we have

$$y = 0.23 \pm (0.77 \cdot 10^{-2} + 0.03 \cdot 10^{-2})$$
$$= 0.23 \pm 0.8 \cdot 10^{-2}.$$

The maximal error bound is very pessimistic if the number of variables is large. In such cases, it is better to use statistical methods and to compute an average value for the error of the result.

In the above presentation, we have had the problem of computing a real valued function of several variables. If the function f is vector valued, i.e.,

$$f = \begin{pmatrix} f_1 \\ f_2 \\ \vdots \\ f_m \end{pmatrix},$$

the methods can be used for error estimates for each component separately.

The relative error of an approximate value is a measure of the information content of the approximation. To measure a quantity, e.g., the distance between two points, with six significant digits usually requires more advanced equipment than measuring the same quantity with two significant digits. When the approximation is used in numerical computations, it is important not to destroy information by using bad algorithms.

Example 2.3.7 The approximations

$$x_1 = 10.123455 \pm 0.5 \cdot 10^{-6},$$

$$x_2 = 10.123789 \pm 0.5 \cdot 10^{-6}$$

both have eight significant digits and relative error less than $0.5 \cdot 10^{-7}$. When we subtract

$$y = x_1 - x_2 = -0.000334 \pm 10^{-6},$$

the approximation of y has a small absolute error, but the relative error can be estimated as

$$\left| \frac{\Delta y}{y} \right| \leq \frac{10^{-6}}{0.000334} < 3 \cdot 10^{-3};$$

the approximation has only two significant digits.

The loss of accuracy that occurs when two almost equal numbers are subtracted is called (subtractive) **cancellation**. Such loss of accuracy can often be avoided by a reformulation into a mathematically equivalent expression.

Example 2.3.8   The second degree equation

$$x^2 - 18x + 1 = 0$$

has the solution $x_{1,2} = 9 \pm \sqrt{80}$. If $\sqrt{80}$ is given with four correct decimals, we get

$$x_1 = 9 + 8.9443 \pm 0.5 \cdot 10^{-4} = 17.9443 \pm 0.5 \cdot 10^{-4},$$

$$x_2 = 9 - 8.9443 \pm 0.5 \cdot 10^{-4} = 0.0557 \pm 0.5 \cdot 10^{-4}.$$

Thus the first approximation has six significant digits, while the second has only three. Cancellation is avoided by rewriting

$$x_2 = \frac{(9 - \sqrt{80})(9 + \sqrt{80})}{9 + \sqrt{80}} = \frac{1}{9 + \sqrt{80}} = \frac{1}{17.9443 \pm 0.5 \cdot 10^{-4}}.$$

Then we get

$$\frac{1}{17.9443} = 0.055728002\ldots.$$

The error propagation rule for division says that the relative error of the approximation of x_2 (coming from the error in the approximation of $\sqrt{80}$) is at most

$$\frac{0.5 \cdot 10^{-4}}{17.9443} < 0.3 \cdot 10^{-5}.$$

The absolute error is less than

$$0.3 \cdot 10^{-5} \cdot 0.05573 < 0.17 \cdot 10^{-6}.$$

If we round the approximation to seven decimals, we get

$$x_2 = 0.0557280 \pm 0.2 \cdot 10^{-6},$$

which has five significant digits.

We give a couple of more examples of reformulations for avoiding cancellation:

$$\sqrt{1+x} - \sqrt{1-x} = \frac{2x}{\sqrt{1+x} + \sqrt{1-x}},$$

$$\log b - \log a = \log \frac{b}{a}.$$

(Note that throughout this book log denotes the natural logarithm.) If it is difficult to find a suitable reformulation of an expression of the form

$$f(x + \epsilon) - f(x),$$

then cancellation can be avoided by using the Taylor expansion; e.g., for small values of x, we have

$$1 - \cos x \approx \frac{x^2}{2!} - \frac{x^4}{4!}.$$

2.4   The Representation of Numbers in Computers

The decimal number system is a **position system** with **base** 10. Since most computers use a position system with another base (e.g., 2 or 16), we shall start this section by recalling how numbers are represented in a position system with arbitrary base β. Let β be a natural number, $\beta \geq 2$; any real number can be written

$$(\pm d_n d_{n-1} \ldots d_2 d_1 d_0.d_{-1} d_{-2} \ldots)_\beta,$$

where $d_n, d_{n-1}, \ldots$ are integers between 0 and $\beta - 1$. The value of such a number is

$$d_n \beta^n + d_{n-1} \beta^{n-1} + \ldots + d_2 \beta^2 + d_1 \beta^1 + d_0 \beta^0 + d_{-1} \beta^{-1} + d_{-2} \beta^{-2} + \ldots.$$

Example 2.4.1

$$(760)_8 = 7 \cdot 8^2 + 6 \cdot 8^1 + 0 \cdot 8^0 = (496)_{10},$$

$$(101.101)_2 = 1 \cdot 2^2 + 0 \cdot 2^1 + 1 \cdot 2^0 + 1 \cdot 2^{-1} + 0 \cdot 2^{-2} + 1 \cdot 2^{-3}$$

$$= (5.625)_{10},$$

$$(0.333\ldots)_{10} = 3 \cdot 10^{-1} + 3 \cdot 10^{-2} + 3 \cdot 10^{-3} \ldots = \frac{1}{3}.$$

The architecture of most computers is based on the principle that data are processed with a fixed amount of information as a unit. Such a unit of information is called a **word**, and the number of digits (usually binary) in a word is called the **word length** of the computer. Common word lengths are

16	(many microprocessors)
32	(IBM 3090, VAX)
64	(supercomputers)

Most microcomputers use 32 bits (binary digits) to represent real numbers. Integers can of course be represented exactly in a computer, provided that the actual word length is large enough for storing all the digits.

On the other hand, most real numbers cannot be represented exactly. Firstly, errors arise when a number is converted from one number system to another. For example,

$$(0.1)_{10} = (0.0001100110011\ldots)_2;$$

thus, the number $(0.1)_{10}$ cannot be represented exactly in a computer with a binary number system.

Secondly, rounding errors arise due to the finite word length of the computer.

How should real numbers be represented in a computer? The first computers (in the forties and early fifties) used **fixed point representation**: for each computation, the user decided how many digits in a computer word were to be used for representing the integer and fractional parts, respectively, of a real number.

Obviously, it is difficult to represent simultaneously both large and small real numbers with this method. Assume, for example, that we have a decimal representation with a word length of six digits, and that we use three digits for decimals. The largest and smallest (positive) numbers that can be represented are 999.999 and 000.001, respectively. Also observe that with fixed point representation, large numbers can be represented with much smaller relative errors than small numbers.

This difficulty can be overcome by representing real numbers in the exponent form that we generally use for very large and small numbers. E.g., we would not write

$$0.00000000789, \qquad 6540000000000$$

but rather

$$7.89 \cdot 10^{-9}, \qquad 6.54 \cdot 10^{12}.$$

This way of representing real numbers is called **floating point representation** (as opposed to fixed point).

Any nonzero real number X in the number system with base β can be written in the form

$$X = M \, \beta^e,$$

where e is an integer, and

$$M = \pm D_0.D_1 D_2 D_3 \dots,$$

$$0 \le D_i < \beta, \qquad i = 0, 1, 2, \dots,$$

$$D_0 \ne 0.$$

M can be a number with infinitely many digits. When a number written in this form is to be stored in a computer, it must be rounded (or chopped; in the sequel, we mostly discuss rounding). Assume that $t + 1$ digits are used for representing M. Then we store the number

$$x = m \, \beta^e,$$

where m is equal to M rounded to $t + 1$ digits, and

$$m = \pm d_0.d_1 d_2 \dots d_t.$$

m is called the **significand** or **mantissa**, and e is called the **exponent**. The digits to the right of the point in m are called the **fraction**. We have already indicated that the significand shall be larger than 1 in magnitude. Now we assume more precisely that the nonzero floating point numbers are normalized so that

$$1 \le |m| < \beta.$$

The left inequality implies that $d_0 \ne 0$. Thus, in floating point representation we avoid storing unnecessary zeros in the beginning of a number. The right inequality implies that the integer part of m has only one digit. The

amount of storage that is reserved for the exponent, determines the range
of numbers that can be represented. The limits of e can be written

$$L \le e \le U,$$

where L and U are negative and positive integers, respectively. If the
result of a computation is a floating point number with $e > U$, then the
computer issues an error signal. This type of error is called **overflow**. The
corresponding error with $e < L$ is called **underflow**. This error is usually
somewhat less serious than overflow.

A set of floating point numbers is uniquely characterized by β (the base),
t (the precision), and (L, U) (the interval of the exponent). In the sequel,
we simply refer to **the floating point system** (β, t, L, U).

A floating point system is the set of **normalized floating point
numbers in the number system with base β, and t digits for
the fraction** (equivalently $t + 1$ digits in the significand), i.e., all
numbers of the form

$$x = m\,\beta^e,$$

where

$$m = \pm d_0.d_1 d_2 \ldots d_t,$$

$$0 \le d_i < \beta, \qquad i = 1, 2, \ldots, t,$$

$$1 \le |m| < \beta,$$

and where the exponent satisfies

$$L \le e \le U.$$

It is important to note that the floating point numbers are not evenly
distributed over the real axis. E.g., assume that $(\beta, t, L, U) = (2, 2, -2, 1)$.
The positive floating point numbers in this system are marked along the
axis in Figure 2.4.2.

Figure 2.4.2   The positive numbers in the floating point system $(2, 2, -2, 1)$.

Some values of (β, t, L, U) are given in Table 2.4.3.

	β	t	L	U
IEEE standard	2	23	-126	127
IBM 3090	16	5	-65	62
CRAY X-MP	2	47	-16385	8190

Table 2.4.3   Examples of floating point number systems.

Note that the computers IBM 3090 and CRAY X-MP in Table 2.4.3 have a different normalization than the one given above. The parameters of these computers have been translated to our system to make it possible to compare the precision of the floating point systems (see next section).

There are **double precision floating point numbers** in several programming languages, e.g., Fortran. Usually, they are implemented so that two computer words are used for storing one number. This gives higher precision and a larger range. Often double precision arithmetic is programmed, whereas single precision arithmetic is built into the hardware in many computers.

We want to emphasize that our notation "the floating point number system (β, t, L, U)" means that the *fraction* occupies t digits. In other literature, floating point numbers are normalized so that $\beta^{-1} \le |m| < 1$, and there t digits are used for the *significand*. Our notation is consistent with the IEEE standard for floating point arithmetic (see Section 2.8).

2.5   Rounding Errors in Floating Point Arithmetic

When we represent numbers in the floating point system (β, t, L, U) rounding errors arise because of the limited precision. We assume that rounding is done and shall derive a bound for the relative error. The corresponding result for chopping will also be given.

Assume that a real number x can be written (exactly)

$$x = m\,\beta^e, \ 1 \le |m| < \beta;$$

further, let $x_r = m_r\,\beta^e$, where m_r is equal to m rounded to $t + 1$ digits. Then we have

$$|m_r - m| \le \frac{1}{2}\beta^{-t},$$

and, as a bound for the absolute error, we get

$$|x_r - x| \le \frac{1}{2}\beta^{-t}\beta^e.$$

We can derive a bound for the relative error:

$$\frac{|x_r - x|}{|x|} \leq \frac{\frac{1}{2}\beta^{-t}\beta^e}{|m|\,\beta^e} = \frac{\frac{1}{2}\beta^{-t}}{|m|}.$$

From the normalization condition $1 \leq |m|$ now follows

$$\frac{|x_r - x|}{|x|} \leq \frac{1}{2}\beta^{-t}.$$

Thus, we have shown

Theorem 2.5.1   The **relative rounding error** in floating point representation can be estimated as

$$\frac{|x_r - x|}{|x|} \leq \mu,$$

where $\mu = \frac{1}{2}\beta^{-t}$ is called the **unit roundoff**.

Observe that the bound for the relative rounding error is independent of the magnitude of the number that is being approximated. Thus, both large and small numbers can be represented with the same relative accuracy.

Sometimes, it is convenient to use the following equivalent notation.

There is an $\epsilon > 0$, that satisfies $|\epsilon| \leq \mu$, such that

$$x_r = x(1 + \epsilon).$$

This notation is especially suitable in the analysis of accumulated rounding errors.

Example 2.5.2   The floating point system $(2, 23, -126, 127)$ has unit roundoff

$$\mu = \frac{1}{2} \cdot 2^{-23} = 2^{-24} \approx 5.96 \cdot 10^{-8}.$$

If we have a binary arithmetic computer with $t + 1$ digits in the significand, how accurate is this computer, expressed in terms of decimal numbers? We must know the answer to this question to be able decide how many decimal digits should be printed in a computed result. The question can also be written: "How many decimal digits correspond to $t + 1$ bits?" or, equivalently, "If the unit roundoff in a binary floating point system is $\mu = 0.5 \cdot 2^{-t}$, how many digits must we have in a decimal system with approximately the same unit roundoff?" Thus, we shall solve the equation

$$\frac{1}{2} \cdot 10^{-s} = \frac{1}{2} \cdot 2^{-t},$$

with respect to s. Taking logarithms, we get

$$s = t \log_{10} 2 \approx 0.3t.$$

Rule of thumb:

t binary digits correspond to $0.3t$ decimal digits.

s decimal digits correspond to $3.3s$ binary digits.

Example 2.5.3 A binary floating point system with $t = 23$ corresponds approximately to a decimal system with $s = 7$, since

$$23 \log_{10} 2 \approx 6.9.$$

The derivations of this section can be made also for floating point systems with chopping. The only difference is that the **unit chopoff** is

$$\mu_c = 2\mu = \beta^{-t}.$$

Floating point arithmetic with chopping is easier to implement than arithmetic with rounding, since all digits to the right of the least significant digit can simply be discarded. In some computers with a large word length, chopping is used since the errors are so small anyway. However, the IEEE standard for floating point arithmetic prescribes that rounding should always be used (see Section 2.8).

2.6   Arithmetic Operations with Floating Point Numbers

The aim of this section is not to describe in detail how floating point arithmetic can be implemented. Rather, we want to show that, under certain assumptions, it is possible to perform floating point operations with good accuracy. This accuracy should then be requested from all implementations.

Since rounding errors arise already when real numbers are stored in floating point, one can hardly expect to be able to perform arithmetic operations without errors. It is enough to note that the result of multiplying two $t + 1$ digit numbers is a number with $2t + 2$ or $2t + 1$ digits. When the assignment statement

$$a := b * c,$$

is executed, where a, b and c are single precision variables, the result must be rounded before it is assigned to the variable a.

Up until 1985, there did not exist a standard for floating point arithmetic, and therefore different computer manufacturers chose their own solutions depending on speed and other factors. In this section, we shall describe a somewhat simplified arithmetic in order to be able to explain the principles of floating point arithmetic without having to go into too much detail. A survey of the floating point standard is given in Section 2.8, and, in Section 2.9, we describe a specific implementation of the standard.

We shall describe how arithmetic operations are performed in a floating point system (β, t, L, U). Rounding is assumed. In the numerical examples, we use the floating point system $(10, 3, -9, 9)$.

Computers have special registers where arithmetic operations are performed. The length of these registers (the number of digits they hold) determine how accurately floating point operations can be performed. We assume here that the arithmetic registers hold $2t + 4$ digits. It is possible to perform the arithmetic operations with the same accuracy (and faster) using shorter registers, but those algorithms are more complicated.

Apart from arithmetic and logical operations, one must be able to perform **shift** operations in order to normalize and to let two floating point numbers have the same exponent. A left shift means that the exponent is decreased:

$$0.031 \cdot 10^1 = 3.100 \cdot 10^{-1}.$$

The first algorithm we discuss is floating point addition (and at the same time subtraction, since $x - y = x + (-y)$).

Assume that

$$x = m_x \beta^{e_x},$$
$$y = m_y \beta^{e_y},$$

and that $x \geq y$. Let $z = \text{fl}[x + y]$ denote the result of the floating point addition. We shift y $e_x - e_y$ positions to the right and then add:

$$1.234 \cdot 10^0 + 4.567 \cdot 10^{-2} = (1.234 + 0.04567) \cdot 10^0 = 1.27967 \cdot 10^0.$$

But, if $e_x - e_y \geq t + 3$, we get $\text{fl}[x + y] = x$. E.g.,

$$1.234 \cdot 10^0 + 5.678 \cdot 10^{-6} = 1.234005678 \cdot 10^0 \doteq 1.234 \cdot 10^0$$

("$\doteq$" should be read "is rounded to"). We get the following addition algorithm.

FLOATING POINT ADDITION $z := x + y$;

    $e_z := e_x$;   ($x \geq y$ is assumed)
    if $e_x - e_y \geq t + 3$ **then** $m_z := m_x$
    else
        $m_y := m_y / \beta^{e_x - e_y}$;   (right shift $e_x - e_y$ positions)
        $m_z := m_x + m_y$;
    endif;

If $e_x - e_y < t + 3$, then m_y can be stored without errors after the shift, since we have assumed that the arithmetic registers can hold $2t + 4$ digits. Likewise, the addition $m_x + m_y$ is performed without errors. In general, the result of these operations is an *unnormalized* floating point number $z = m_z \beta^{e_z}$. It may happen that $|m_z| \geq \beta$ (e.g., $5.678 \cdot 10^0 + 4.567 \cdot 10^0 = 10.245 \cdot 10^0$), or $|m_z| < 1$ (e.g., $5.678 \cdot 10^0 + (-5.600 \cdot 10^0) = 0.078 \cdot 10^0$). Further, the significand must be rounded to $t + 1$ digits.

The following normalization algorithm takes an unnormalized floating point number $m \beta^e$ as input and gives a normalized floating point number x as output.

NORMALIZATION

    if $|m| \geq \beta$ **then**
        $m := m / \beta$;   (right shift one position)
        $e := e + 1$;
    else
        while $|m| < 1$ **do**
            $m := m * \beta$;   (left shift one position)
            $e := e - 1$;
    endif;
    Round m to $t + 1$ digits;
    if $|m| = \beta$ **then**

$$m := m/\beta; \qquad \text{(right shift one position)}$$
$$e := e + 1;$$
endif;
if $e > U$ **then** exponent overflow;   (error signal)
if $e < L$ **then**
      exponent underflow;   ("soft" error signal)
$$x := 0;$$
else
$$x := m \, \beta^e;$$
endif   (the normal case)

Rounding to $t + 1$ digits may give an unnormalized result:

$$9.9995 \cdot 10^3 \doteq 10.000 \cdot 10^3 = 1.000 \cdot 10^4.$$

This explains the if-statement after the rounding.
   The multiplication and division algorithms are simple.

FLOATING POINT MULTIPLICATION $z := x * y$;

$$e_z := e_x + e_y;$$
$$m_z := m_x * m_y;$$
Normalize;

FLOATING POINT DIVISION $z := x/y$;

   if $y = 0$ **then** division by zero   (error signal)
   else
$$e_z := e_x - e_y;$$
$$m_z := m_x/m_y;$$
Normalize;
   endif

We have assumed that the arithmetic registers hold $2t + 4$ digits. This means that the results of addition and multiplication before normalization and rounding are exact. Therefore, the only error in these operations is the rounding error. A more careful analysis of the division algorithm shows that the division of the significands can be performed without errors to $2t + 4$ digits. Therefore, we have the same fundamental error estimate for the arithmetic operations as for floating point representation (Theorem 2.5.1).

Theorem 2.6.1   Let $\odot$ denote any of the operations $+, -, *$ and $/$. Then (if $x \odot y \neq 0$, and provided that the registers are as assumed above)

$$\left| \frac{x \odot y - \mathrm{fl}[x \odot y]}{x \odot y} \right| \leq \mu,$$

or, equivalently,

$$\mathrm{fl}[x \odot y] = (x \odot y)(1 + \epsilon),$$

for some $\epsilon > 0$ that satisfies $|\epsilon| \leq \mu$ ($\mu = \frac{1}{2}\beta^{-t}$ is the unit roundoff).

It can be shown that the theorem is valid even if the registers hold $t + 4$ digits only, provided that the algorithms are modified accordingly.

A consequence of the errors in floating point arithmetic is that the usual mathematical laws are no longer valid. E.g., the associative law for addition and multiplication does not hold:

$$\mathrm{fl}[\mathrm{fl}[a + b] + c] \neq \mathrm{fl}[a + \mathrm{fl}[b + c]].$$

Example 2.6.2   Let $a = 9.876 \cdot 10^4$, $b = -9.880 \cdot 10^4$ and $c = 3.456 \cdot 10^1$ in the floating point system $(10, 3, -9, 9)$. We get

$$\mathrm{fl}[\mathrm{fl}[a + b] + c] = \mathrm{fl}[-4.000 \cdot 10^1 + 3.456 \cdot 10^1] = -5.440 \cdot 10^0,$$

whereas

$$\mathrm{fl}[a + \mathrm{fl}[b + c]] = \mathrm{fl}[9.876 \cdot 10^4 - 9.877 \cdot 10^4] = -1.000 \cdot 10^1.$$

Another consequence of Theorem 2.6.1 is that it is seldom meaningful to test for equality between floating point numbers. If, in the program statement below, x and y are declared **real**,

        if $x = y$ **then** ...,

and are the results of earlier computations, then the probability that the boolean expression is true is very small. Instead one should write

        if abs(x-y)< eps **then** ...,

for some small constant eps.

2.7   Accumulated Errors

We shall take the computation of a sum

$$S_n = \sum_{k=1}^{n} x_k$$

as an example of error accumulation in repeated floating point operations. It is practical to use the error estimate for addition in the form

$$\text{fl}[a + b] = (a + b)(1 + \epsilon),$$

for some ϵ with $|\epsilon| \leq \mu$. If we compute the sum in the natural order ($\hat{S}_i$ denotes a computed partial sum)

$$\hat{S}_1 := x_1,$$
$$\hat{S}_i := \text{fl}[\hat{S}_{i-1} + x_i], \qquad i = 2, 3, \ldots, n,$$

we get

$$\hat{S}_i = (\hat{S}_{i-1} + x_i)(1 + \epsilon_i), \qquad i = 2, 3, \ldots, n.$$

A simple induction argument gives (compute, e.g., $\hat{S}_2$ and $\hat{S}_3$ explicitly)

$$\hat{S}_n = \hat{x}_1 + \hat{x}_2 + \ldots + \hat{x}_{n-1} + \hat{x}_n,$$

where

$$\hat{x}_1 = x_1(1 + \epsilon_2)(1 + \epsilon_3) \cdots (1 + \epsilon_n),$$
$$\hat{x}_2 = x_2(1 + \epsilon_2)(1 + \epsilon_3) \cdots (1 + \epsilon_n),$$

and, in general,

$$\hat{x}_i = x_i(1 + \epsilon_i)(1 + \epsilon_{i+1}) \cdots (1 + \epsilon_n), \qquad i = 2, 3, \ldots, n.$$

To be able to obtain practical error estimates, we need the following lemma, which we give without proof.

Lemma 2.7.1   Let $\epsilon_1, \epsilon_2, \ldots, \epsilon_n$ be numbers satisfying $|\epsilon_i| \leq \mu$, $i = 1, 2, \ldots, n$, and assume that $n\mu < 0.1$. Then there is a number δ_n such that

$$(1 + \epsilon_1)(1 + \epsilon_2) \cdots (1 + \epsilon_n) = 1 + \delta_n,$$

and

$$|\delta_n| \leq 1.06n\mu.$$

We can now derive two types of results, which give error estimates for summation in floating point arithmetic.

Theorem 2.7.2 Forward analysis. If $n\mu < 0.1$, then the error in the computed sum can be estimated as

$$\left|\hat{S}_n - S_n\right| \leq |x_1|\,|\delta_{n-1}| + |x_2|\,|\delta_{n-1}| + \ldots + |x_n|\,|\delta_1|,$$

where
$$|\delta_i| \leq i \cdot 1.06\mu, \qquad i = 1, 2, \ldots, n-1.$$

Proof. According to Lemma 2.7.1 we can write

$$\hat{S}_n = x_1(1 + \delta_{n-1}) + x_2(1 + \delta_{n-1}) + \ldots + x_n(1 + \delta_1),$$

where δ_i satisfies the inequality in the theorem. Subtract S_n and use the triangle inequality. ∎

Forward analysis is the type of error analysis that we have used at the beginning of this chapter. However, it is difficult to use this method to analyze such a fundamental algorithm as Gaussian elimination for the solution of linear systems of equations. In the 1950s, J.H. Wilkinson was the first to make a correct error analysis of this algorithm using **backward error analysis**.

In backward analysis, one shows that the approximate solution $\hat{S}$ that has been computed for the problem P is the *exact solution* of a perturbed problem $\hat{P}$. We cite the following description of the aim of backward analysis from a book on numerical linear algebra:

> "The objective of backward error analysis is to stop worrying about whether one has the "exact" answer, because this is not a well-defined thing in most real-world situations. What one wants is to find an answer which is the true mathematical solution to a problem which is within the domain of uncertainty of the original problem. Any result that does this must be acceptable as an answer to the problem, at least with the philosophy of backward error analysis." (J. R. Rice, *Matrix computations and mathematical software*, McGraw–Hill, New York, 1981.)

In the summation example, we can formulate

Theorem 2.7.3   Backward analysis. Assume that $n\mu < 0.1$. Then, we have

$$\hat{S}_n = \hat{x}_1 + \hat{x}_2 + \ldots + \hat{x}_n,$$

where

$$\hat{x}_1 = x_1(1 + \delta_{n-1}),$$

$$\hat{x}_i = x_i(1 + \delta_{n-i+1}), \qquad i = 2, 3, \ldots, n,$$

$$|\delta_i| \leq i \cdot 1.06\mu, \qquad i = 1, 2, \ldots, n-1.$$

One important conclusion can be drawn from the error estimates in Theorems 2.7.2 and 2.7.3. Let us rewrite the error estimate of the forward analysis as

$$\left|\hat{S}_n - S_n\right| \leq [(n-1)\,|x_1| + (n-1)\,|x_2| + (n-2)\,|x_3| + \ldots$$
$$\ldots + 2\,|x_{n-1}| + |x_n|]1.06\mu.$$

We immediately see that in order to minimize the error bound, we should add the terms in increasing order, since the first terms in the summation have the largest multipliers in the error estimate.

Example 2.7.4   Let

$$x_1 = 1.234 \cdot 10^1,$$
$$x_2 = 3.453 \cdot 10^0,$$
$$x_3 = 3.441 \cdot 10^{-2},$$
$$x_4 = 4.667 \cdot 10^{-3},$$
$$x_5 = 9.876 \cdot 10^{-4}.$$

If we perform the summation in decreasing order in the floating point system $(10, 3, -9, 9)$ (with rounding), we get the sum $1.582 \cdot 10^1$. Summation in increasing order gives $1.583 \cdot 10^1$. The correct result rounded to six decimals is

$$S_5 = 1.583306 \cdot 10^1.$$

Similarly, a relatively large error arises when a slowly converging series is summed in decreasing order.

Example 2.7.5 We computed the sum

$$\sum_{n=1}^{30000} \frac{1}{n^2}$$

in the floating point system $(2, 22, -129, 126)$ with rounding, in decreasing and increasing order, and we obtained the results 1.644860 and 1.644901, respectively.

We must point out, however, that the major part of the difference is due to the fact that in the first case none of the approximately 18000 last additions contributes to the result even in the last bit, since here

$$\text{fl}[\text{S} + \frac{1}{\text{n}^2}] = \text{S},$$

where S is the summation variable.

2.8 The IEEE Standard for Floating Point Arithmetic

Above all, it was the development of microcomputers that made it necessary to standardize floating point arithmetic. One of the aims was to facilitate portability, so that it will be possible to run the same program without changes on different computers. If two computers conform to the standard, then the execution of the same program on the two computers should give identical results. This is not the case for computers with non-standardized arithmetic.

A proposal for a standard for binary floating point arithmetic was presented in 1979. Some changes were made, and the standard was adopted in 1985. It has been implemented in several microcomputers (INTEL 8087 and Motorola 68881 were among the first). We shall present the standard without going into too much detail.

The standard defines four floating point formats in two groups, **basic** and **extended**, each with two widths, **single** and **double**. The organization of the single precision basic format is shown in Figure 2.8.1.

Figure 2.8.1 Basic format, single precision.

The components are the sign σ (one bit), the exponent e (eight bits), and the fraction f (23 bits), altogether 32 bits.

The value v of a floating point number x is

    a. If $e = 255$ and $f \neq 0$, then v is NaN (Not-a-Number, see below).
    b. If $e = 255$ and $f = 0$, then $v = (-1)^\sigma \infty$.
    c. If $0 < e < 255$, then $v = (-1)^\sigma (1.f) 2^{e-127}$.
    d. If $e = 0$ and $f \neq 0$, then $v = (-1)^\sigma (0.f) 2^{-126}$ (unnormalized).
    e. If $e = 0$ and $f = 0$, then $v = (-1)^\sigma 0$ (zero).

The normal case is c. We see that the leading bit of the significand, i.e., the one to the left of the binary point, is not stored, since due to the normalization ($1 \leq t < 2$) it is known always to be equal to one. Thus, one extra bit is gained for the fraction. Further, the exponent is not treated as we assumed in Section 2.4; rather, it is stored in **biased** form.

Example 2.8.2   The numbers 1, 4.125 and $-0.09375 = -\frac{3}{32}$ are stored

0	0 1 1 1 1 1 1 1	0 0 0 0 0 0 0	$\cdots$	0

0	1 0 0 0 0 0 0 1	0 0 0 0 1 0 0	$\cdots$	0

1	0 1 1 1 1 0 1 1	1 0 0 0 0 0 0	$\cdots$	0

The largest and smallest positive numbers that can be represented are $2^{127}(2 - 2^{-23})$ and 2^{-126}, respectively.

One reason for introducing the entity NaN (Not-a-Number) is to make debugging easier. When an invalid operation is performed (e.g., 0/0), or when an uninitialized variable is used, the result shall be a NaN, which contains debugging information. In many cases, it is not necessary to stop the execution of a program when an invalid operation has been performed.

For a long time it has been common practice to put the result equal to zero when an operation gave underflow. The IEEE standard defines the option **"gradual underflow"** (case d).

The details of the extended single precision format are left to the implementer, but the minimal requirements are a sign bit σ, at least 32 bits in the significand, an exponent e that may be biased and must satisfy

$$L \leq e \leq U, \qquad L \leq -1023, \qquad U \geq 1024.$$

The basic double precision format is analogous to the single precision format. The sign σ is one bit, the exponent e is 11 bits, and the fraction f is 52 bits, altogether 64 bits. In the extended double precision format, at least 63 bits are used for the fraction.

Implementations of the standard shall provide the addition, subtraction, multiplication, division and square root operations, as well as binary-decimal conversions. Except for the conversions, all operations shall give a result that is equal to the rounded result of the corresponding operation correct to infinite precision. This means that Theorem 2.6.1 is valid.

The standard rounding mode is the one we have described earlier in this chapter. In particular, rounding to even is prescribed in the case when two values are equally near.

The extended formats can be used both for avoiding overflow and to give better accuracy (note however that high level languages as a rule do not give the programmer this option).

Example 2.8.3   The computation of

$$\text{sq} := \sqrt{x_1^2 + x_2^2},$$

can give overflow even if the result can be represented (see Exercise 11). If the compiler uses extended format for the computed squares, then overflow cannot occur.

Example 2.8.4   The length of a vector

$$l = \left(\sum_{i=1}^{n} x_i^2 \right)^{1/2}$$

can be computed by the following program.

```
Extended real s;
s := 0;
for i := 1 to n do s := s + x_i * x_i;
l := √s;
```

If l can be represented (e.g., in single precision) overflow cannot occur. Further, since the significand of the extended format has more digits, l will be computed more accurately than in the case when s is a basic format variable. In fact, if n is not too large, l can even be equal to the exact result rounded to the basic format.

2.9   The Floating Point Processor INTEL 8087

To illustrate further the IEEE standard for floating point arithmetic we here describe the implementation of the standard in a commercially available microprocessor. The main reason for choosing the INTEL 8087 is that it has been documented in a much more readable form than is usual for most computer products (see the list of references at the end of the chapter).

INTEL 8087 is a powerful microprocessor for floating point arithmetic (and even more so its descendants 80287 and 80387, which are similar in their basic architecture). It cannot be used on its own, but only connected to a 8086 or 8088 processor, and it is used for executing floating point instructions in hardware. Thus, 8087 can be used in computers based on 8086, e.g., IBM PC and compatibles, and the speed of execution for numerical computations is increased dramatically when the 8087 is used. It has been estimated that linear systems of equations can be solved more than 20 times faster with the floating point processor than without.

It is not possible to give more than an introduction to the 8087, which, when constructed, was INTEL's most advanced chip so far. We are going to describe enough to see how simple computations like summation can be implemented efficiently both with respect to speed and accuracy.

As prescribed in the floating point standard, the 8087 has single and double precision floating point numbers, with 32 and 64 bits, respectively. But there is only one extended format, which is common to single and double precision. Eighty bits are used for extended precision floating point numbers.

Figure 2.9.1   Extended floating point format.

We here have a sign bit σ, an exponent e with 15 bits, and a significand with 64 bits. Note that also the bit to the left of the binary point is stored. With this organization of the 80 bits, the representation complies exactly to the requirements of the standard for the extended double precision format.

The 8087 has eight arithmetic registers, each with 80 bits (to be able to hold a extended precision floating point number). The registers are organized in a *stack*. There is a three bit status register, which holds a pointer ST to the top register on the stack.

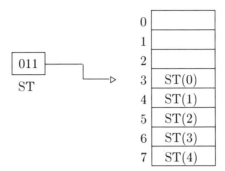

Figure 2.9.2   Arithmetic registers organized in a stack.

In Figure 2.9.2, we see the notation used for the different registers (the notation is from the 8086/8087 assembler language). To make a reference to the top of the stack one can use either ST(0) or ST. In the figure the registers 0,1 and 2 are empty. When a number is loaded from primary memory to an arithmetic register it is put on the top of the stack.

We start from the situation in Figure 2.9.2. When the instruction

FLD   X ;   "Push and load"

has been executed the stack is as follows:

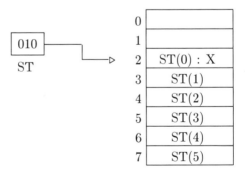

Figure 2.9.3

When the instruction FLD X is executed, ST is decreased by one, the variable X is loaded from memory and is put on the top of the stack. The contents of the remaining registers are the same, but the names are changed.

The number that is fetched from memory can be a single precision, double precision or extended format number. Before it is placed in the register, it is converted to extended format. All arithmetic operations are performed in extended format. The assumptions of Theorem 2.6.1 are therefore valid for the single precision format, and the operation $z := x \odot y$, where $\odot$ means $+, -, *,$ or $/$, can be performed with the accuracy stated in the theorem. Also, the double precision operations can be performed to full accuracy, since the registers are long enough (cf. the remark after Theorem 2.6.1). The movement of data from the stack to primary memory is performed with any of the instructions

FST   X ;   "Store"
FSTP   X ;   "Store and pop"

The first one means that the number on the top of the stack is stored in memory in the location reserved for the variable X, after the number has been converted to the format specified in the declaration of X. FSTP means that after the number has been moved to memory, ST is increased by one. Thus, the top of the stack is moved one step down.

Some syntactical variations of the addition instruction are illustrated in Figure 2.9.4.

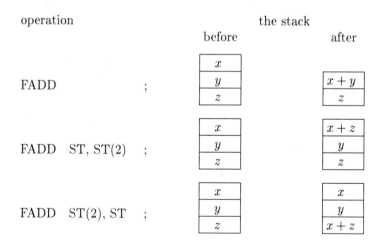

Figure 2.9.4   The instruction FADD.

We now give two examples how the extended format can be used to obtain high accuracy in computations. The first is the computation of the sum

$$S = \sum_{i=1}^{n} x_i.$$

In Pascal-like code, we can write the summation

$$s := 0;$$
$$\textbf{for } i := 1 \textbf{ to } n \textbf{ do } s := s + x_i;$$

When generating machine code for this program segment, it is essential to keep the variable s on the stack during the whole loop. Then, s is represented in extended format throughout the computation, and the dominating error is the rounding error when the result is stored in memory. The segment can be coded

```
        FLDZ              ;   put 0 on the stack
        MOV    CX, N-1     ;   initialize the CX-register in 8086 for
                              the loop instruction.
        XOR    SI, SI      ;   zero the index register.
        ADD    SI,1        ;   put one in the index register.
START:
        FADD   ST, X[SI]   ;   add x_i.
        ADD    SI, 4       ;   increase the index register by 4.
                              (It is assumed that x_i is in single
                              precision (4 bytes)).
        LOOPNZ   START ;       repeat n times.
        FSTP   S           ;   store s.
```

The following program segment

$$s := 0;$$
$$t := 0;$$
$$\textbf{for } i := 1 \textbf{ to } \text{n} \textbf{ do}$$
$$\quad s := s + x_i;$$
$$\quad t := t + x_i * y_i;$$

is just a little more complicated to code for 8087.

```
        FLDZ              ;   put 0 in the two
        FLDZ              ;   top registers of the stack.
        MOV CX, N-1       ;   initialize the CX register for the loop.
        XOR SI, SI        ;
        ADD SI, 1         ;
START:
        FLD X[SI]         ;   put x_i on the top of the stack.
        FADD ST(2), ST    ;   add x_i to s.
        FMUL ST, Y[SI]    ;   multiply x_i * y_i.
```

FADD   ;   add $x_y * y_i$ to t.
ADD SI, 4   ;
LOOPNZ START ;
FSTP T   ;   store t.
FSTP S   ;   store s.

In Figure 2.9.5, we trace the contents of the stack during the execution of the statements of the loop.

START:

t
s

FLD X[SI]

x_i
t
s

FADD ST(2), ST

x_i
t
$s + x_i$

FMUL ST, Y[SI]

$x_i \cdot y_i$
t
$s + x_i$

FADD

$t + x_i \cdot y_i$
$s + x_i$

Figure 2.9.5

In the two preceding examples it is rather evident that the summation variables s and t should not be stored in memory during the summation. Note however that, e.g., the semantics of Fortran *prescribe* them to be stored every time they are assigned a new value inside the loop. Thus, a Fortran compiler cannot take advantage of the extended precision.

There are other cases when accuracy and speed can be gained by keeping results on the stack. Consider the following two assignment statements, which both have the same subexpression:

$$c := b + \sqrt{a + b}; \qquad d := a + \sqrt{a + b};$$

Since neither a nor b is changed between the square root computations, it is enough to compute $\sqrt{a + b}$ once and keep it on the stack, thus saving time.

The computation is performed by the following code.

```
FLD A          ;    put a on the top of the stack.
FLD B          ;    put b on the top of the stack.
FLD A          ;    put a on the top of the stack.
FADD ST,ST(1)  ;    a + b on the top.
FSQRT          ;    √a + b on the top.
FADD ST(2),ST  ;    a + √a + b in ST(2).
FADD           ;    b + √a + b on the top.
FSTP C         ;    store c.
FSTP D         ;    store d.
```

As an exercise, trace the contents of the stack during the execution of the code.

Code optimization by elimination of common expressions is done by several Fortran compilers. A sequence of statements is searched for common expressions. It may happen that such expressions cannot be kept on the stack between the execution of the actual statements (other complicated computations may require the whole stack). By storing the results temporarily in memory in extended format, the same accuracy can be obtained and time can be saved.

The methods for gaining speed and accuracy that have been described in this section are applicable for any processor that implements the IEEE standard.

2.10 Pipelined Floating Point Operations

In this section, we discuss how floating point operations are efficiently executed on many modern computers using pipelined floating point arithmetic. This concept is described with addition as an example.

We start by recalling that, in a digital computer, events are taking place at *discrete* points in time t_ν, $\nu = 1, 2, \ldots$. The quantity $t_c = t_{\nu+1} - t_\nu$ is called the **clock period** or **cycle time**. The clock period of a personal computer is of the order 100 ns (1 ns $= 10^{-9}$ s), and the fastest supercomputers (1990) have a clock period of a couple of nanoseconds. It is the hardware technology and also the physical size of the central processing unit (CPU) that determine how short the clock period can be made. The shorter the clock period, the faster the computer.

To introduce the concept of a pipeline, consider an assembly line for making cars, and assume, for simplicity, that the line has only three stages, which take equally long (one time unit).

The normal operation of such an assembly line is to input enough material for one car into the procedure every time unit, so that the workers are active all the time and produce one car every time unit.

Next, consider floating point addition. From Section 2.6 and the following example, we see that floating point addition can be divided into stages:

$$1.234 \cdot 10^0 + 4.567 \cdot 10^{-2} = (1.234 + 0.04567) \cdot 10^0 = 1.27967 \cdot 10^0 \doteq 1.280 \cdot 10^0.$$

For simplicity, we assume here three stages:

Assume that each stage takes one clock period. When the sum of two vectors $z := x + y$ (i.e., $z_i := x_i + y_i$, $i = 1, 2, \ldots, n$) is computed in a computer with pipelined floating point arithmetic, then the addition unit is operated like a car assembly line, a **pipeline**: every clock period a pair of operands is input, and (after an initial startup time) a result is output every clock period.

At a certain point in the computation, the operands have progressed through the pipeline as illustrated below.

| $x_8 \rightarrow$ | x_7 | | x_6 | | x_5 | |
| $y_8 \rightarrow$ | y_7 | $\rightarrow$ | y_6 | $\rightarrow$ | y_5 | $\rightarrow z_4$ |

In the above example, the computation in **vector mode** of the sum of the two vectors takes $3 + n$ clock periods, where n is the vector length. The corresponding computation in **scalar mode** (without pipelining) would take $3n$ clock periods.

Other vector operations that can be pipelined are (componentwise) multiplication $z_i := x_i * y_i$, $i = 1, 2, \ldots, n$, and division $z_i := y_i / x_i$, $i = 1, 2, \ldots, n$. It is also common that two arithmetic functional units can be **chained** together to form a single pipeline. By chaining the multiplication and addition units, the computation of the vector operation

$$y := y + \alpha * x;$$

where α is a scalar, can be performed in vector mode, i.e., so that one result is output every clock period. (This operation is often referred to as the **Saxpy** operation.)

Often computers with pipelined arithmetic have **vector instructions**: e.g., the vector sum $z := x + y$ is computed by one single machine instruction.

Let l denote the startup and unit time for a certain vector operation, i.e. the time to set up the vector instruction plus the time for the first pair of operands to pass through the pipeline. Then, the time to perform that operation on vectors with length n is

$$t = (l + n)t_c,$$

where t_c is the clock period. The rate of producing n results is

$$r_n = \frac{n}{(l + n)t_c}.$$

The **maximum** (or **asymptotic**) **rate** is obtained by letting n tend to infinity:

$$r_\infty = \frac{1}{t_c}.$$

Another interesting parameter is the **half performance length** $n_{1/2}$: the vector length required to achieve half the maximum performance. This can be determined from the equation

$$\frac{n}{(l + n)t_c} = \frac{r_\infty}{2} = \frac{1}{2t_c},$$

which gives

$$n_{1/2} = l.$$

Example 2.10.1 The CRAY X-MP has clock period $t_c = 8.5$ ns. This gives an asymptotic rate for, e.g., vector (componentwise) multiplication of $r_\infty = 1/8.5 \cdot 10^{-9} \approx 117$ Mflops/s (1 Mflop $= 10^6$ floating point operations).

The startup time l for multiplication is $l = 9$ clock periods. Therefore, the half performance length is $n_{1/2} = 9$. This indicates that the CRAY X-MP is very fast for short vectors also.

For the chained SAXPY operation $y := y + \alpha * x$, the asymptotic rate is $r_\infty \approx 234$ Mflops/s, since here the result of two arithmetic operations is output every clock period.

It should be emphasized that the values of these parameters are only *theoretical*. In practice, one has to take into account the time for memory accesses. The measured values are $r_\infty = 70$, $n_{1/2} = 53$ for vector multiplication, and $r_\infty = 148$, $n_{1/2} = 60$ for the SAXPY operation (from R.W. Hockney, C.R. Jesshope, *Parallel Computers 2*, Adam Hilger, Bristol, 1988).

Exercises

1. Show that if u is correctly rounded to s significant digits, then we have

$$\frac{|\Delta u|}{|u|} \le \frac{1}{2}\beta^{-s+1},$$

   where β is the base of the position system.

2. How accurately do we need to know an approximation of π to be able to compute $\sqrt{\pi}$ with four correct decimals?

3. Derive the error propagation formula for division.

4. Let $y = \log x$. Derive the error propagation formula for this function. Use the result to give an error propagation formula for $f(x_1, x_2, x_3) = x_1^{\alpha_1} x_2^{\alpha_2} x_3^{\alpha_3}$. This technique is sometimes called logarithmic differentiation.

5. Compute the focal distance f of a lens using the formula

$$\frac{1}{f} = \frac{1}{a} + \frac{1}{b},$$

   where $a = 32 \pm 1$ mm and $b = 46 \pm 1$ mm. Give an error estimate.

6. When I lie on the beach, I can just see the top of a factory chimney across the water. On my road map, I find that the factory is on the other side of the bay 25 ± 1 km away. I recall that the radius of the earth is 6366 ± 10 km. Compute the height of the chimney and estimate the error.

   Hint: Elementary geometry gives

$$h = \frac{r(1 - \cos\alpha)}{\cos\alpha}, \quad \alpha = \frac{a}{r},$$

where a is the distance to the factory and r is the radius of the earth.

7. Let f be a function from R^n to R^m, and assume that we want to compute $f(\bar{a})$, where the vector $\bar{a}$ is an approximation of a. Show that the general error propagation formula applied to each component in f leads to

$$\Delta f \approx J \, \Delta a,$$

where J is an $m \times n$ matrix with elements

$$(J)_{ij} = \frac{\partial f_i}{\partial a_j}.$$

8. Use Taylor expansion to avoid cancellation in the following expression
   a) $e^x - e^{-x}$,　x close to 0;
   use a reformulation to avoid cancellation in the following expressions
   b) $\sin x - \cos x$,　x close to $\pi/4$,
   c) $1 - \cos x$,　x close to 0,
   d) $(\sqrt{1+x^2} - \sqrt{1-x^2})^{-1}$,　x close to 0.

9. Show that if f is a normalized floating point number in a floating point system (β, t, L, U), then $r \le |f| \le R$, where

$$r = \beta^L,$$

$$R = \beta^U (\beta - \beta^{-t}).$$

10. Show that $\text{fl}[1 + x] = 1$ for all $x \in [0, \mu]$ and that $\text{fl}[1 + x] > 1$ for $x > \mu$ (μ is the unit roundoff of the floating point system).

11. Show that the computation of $\text{sq} := \sqrt{x_1^2 + x_2^2}$ can give overflow even if the result sq can be represented in the floating point system (e.g., take $x_1 = x_2 = 0.8 \cdot 10^5$ in the system $(10, 4, -9, 9)$). Rewrite the computation so that overflow is avoided for all data x_1, x_2 such that the result sq can be represented.

12. Assume that $n\mu < 0.1$ and $|\epsilon_i| \le \mu$, $i = 1, 2, \ldots, n$. Show that

$$|(1 + \epsilon_1)(1 + \epsilon_2) \cdots (1 + \epsilon_n)| \le 1 + 1.06n\mu.$$

Hint: Use $(1 + x)^n \le e^{nx}$ and make a series expansion.

13. Let $S_n = \sum_{i=1}^{n} x_i y_i$. Show that

$$\left| \hat{S}_n - S_n \right| \le \sum_{i=1}^{n} |(n - i + 2) x_i y_i| \ 1.06\mu.$$

(Cf. Theorem 2.7.2.)

14. Assume that $n = 2^k$, and that we compute the sum $S_n = \sum_{i=1}^{n} x_i$ in the order illustrated in the figure.

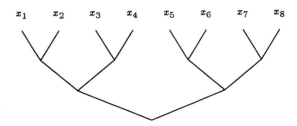

Derive the forward and backward error estimates (see Theorems 2.7.2 and 2.7.3) for this computation.

15. The second degree polynomial $p(x) = ax^2 + bx + c$ is evaluated in a floating point system using Horner's scheme (see Chapter 4). Show that the computed value $\hat{p}(x)$ satisfies

$$|\hat{p}(x) - p(x)| \le (4|ax^2| + 3|bx| + |c|)\mu,$$

where μ is the unit roundoff. (Terms that are $O(\mu^2)$ can be discarded.)

References

The historical development of the representation of numbers is a fascinating chapter in the cultural history of mankind. A nice survey is given in

D. E. Knuth, *The art of computer programming*, Volume 2 /Semi-numerical algorithms, Second edition, Addison–Wesley, Reading, Massachusetts, 1981.

One is tempted to believe that the binary number system is a fruit of the development of computers. As a matter of fact, several mathematicians in the 17th and 18th centuries used binary representation for number theoretical research. Knuth's book gives a good presentation of floating point arithmetic, both single, double and multiple precision. There is also a description (p. 201) of how to perform single precision arithmetic so that Theorem 2.6.1 holds, even if the arithmetic registers can take only $t + 4$ digits.

The first summary of error analysis for floating point arithmetic was given in

J. H. Wilkinson, *Rounding errors in algebraic processes*, Prentice–Hall, Englewood Cliffs, New Jersey, 1963.

The standard for floating point arithmetic has been published by the IEEE

ANSI/IEEE 754, *Binary Floating–Point Arithmetic*, The Institute of Electrical and Electronics Engineers, Inc., 345 East 47th Street, New York, New York 10017.

An implementation of the standard is described in

J. F. Palmer and S. P. Morse, *The 8087 Primer*, Wiley and Sons, New York, 1984.

3   Function Evaluation

3.1   Introduction

When numerical computations are performed, one often uses the elementary functions of mathematics (in connection with computers they are often called **standard functions**). This is the case both for hand calculations and computations with computers. It is necessary to be able to compute these functions easily and efficiently. In hand calculations, it is often practical to use Taylor or Maclaurin expansions of the elementary functions. Series expansions of non-elementary functions are sometimes used, e.g., for the solution of differential equations.

When series expansions are used for numerical computations, the sum of the series is approximated by a partial sum. Then one must estimate the truncation error, the remainder term. In this chapter, we describe the simplest remainder term estimates.

Computer implementations of standard functions are often based on the approximation of the function by a polynomial or a rational function. Here it is important, on the one hand, that the approximation satisfies the accuracy requirements, and, on the other, that it can be evaluated very rapidly. We briefly describe how a standard function can be implemented. Then we discuss some numerical aspects of range reduction, i.e., how mathematical identities can be used to reduce the argument to an interval close to the origin. Finally, we describe a relatively new method, the Cordic algorithm, for implementing trigonometric and other standard functions.

3.2   Numerical Remainder Term Estimates

Let $S = \sum_{n=1}^{\infty} a_n$ be a convergent series (we shall also use S as a notation for the sum of the series). We assume that we cannot compute the sum of the series with analytic methods, and that we are confined to numerical approximations.

The **partial sum** S_N is defined as

$$S_N = \sum_{n=1}^{N} a_n.$$

We are going to use the partial sum as an approximation of S. The corresponding truncation error, the **remainder term** or **tail**, is defined as

$$R_N = S - S_N = \sum_{n=N+1}^{\infty} a_n.$$

When estimating the truncation error, we look for a bound on $|R_N|$.

First, consider the case when the series is alternating, and the absolute value of the terms tends monotonically to zero:

$$a_n a_{n+1} < 0, \quad |a_n| > |a_{n+1}|, \quad n = N, N+1, \ldots, \qquad \lim_{n \to \infty} |a_n| = 0.$$

In Figure 3.2.1, we have illustrated the partial sums as a function of N.

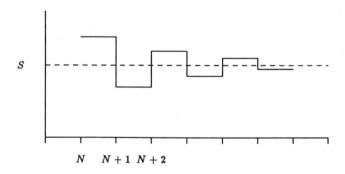

S

$N \quad N+1 \; N+2$

Figure 3.2.1   The partial sums of an alternating series.

It is immediately seen from the figure that

$$S_N \geq S, \quad S_{N+1} \leq S;$$

the partial sums $S_N, S_{N+2}, S_{N+4}, \ldots$ form a monotone, bounded sequence. Thus, the series is convergent. We also see that the tail can be approximated by the first neglected term.

The remainder term of an **alternating series** can be estimated by the first neglected term, i.e.,

$$|R_N| \leq |a_{N+1}|.$$

Example 3.2.2 Let

$$S = \sum_{n=1}^{\infty} \frac{(-1)^{n+1}}{n^2}.$$

We shall compute S with three correct decimals. How many terms are needed in the partial sum?

We have

$$R_N \leq \frac{1}{(N+1)^2} \leq 0.5 \cdot 10^{-3},$$

which gives

$$N \geq \sqrt{2000} - 1 \approx 43.7.$$

Hence, we shall take 44 terms in the partial sum.

We can easily get a good approximation of S with somewhat less work. From Figure 3.2.1, we see that if instead we approximate S by $S_N + \frac{1}{2}a_{N+1}$, then the error estimate is only half as large:

$$S = \left(S_N + \frac{1}{2}a_{N+1}\right) \pm \frac{1}{2}|a_{N+1}|.$$

With this improvement, we get the inequality

$$\frac{1}{2(N+1)^2} \leq 0.5 \cdot 10^{-3},$$

which leads to

$$N \geq \sqrt{1000} - 1 \approx 30.6.$$

Here it is enough to take 31 terms.

We next estimate the remainder term of a positive series, i.e., a series with positive terms. Thus, we have $S = \sum_{n=1}^{\infty} a_n$, where $a_n \geq 0$. In some cases, such a series can be written

$$S = \sum_{n=1}^{\infty} f(n),$$

for some simple function $f(x)$, which is positive and monotonically decreasing for x large enough. If a primitive function F of f is known, then we can use the following relation between a series and an integral:

$$R_N = \sum_{n=N+1}^{\infty} f(n) \leq \int_{N}^{\infty} f(x)\,dx = F(N)$$

(we have of course assumed that $F(x)$ tends to zero as x tends to infinity). It is seen from Figure 3.2.3 that the inequality holds.

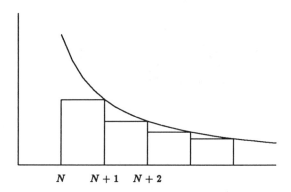

Figure 3.2.3   Estimation of a series by an integral.

Estimation by an integral. The remainder term of the series

$$S = \sum_{n=1}^{\infty} f(n),$$

where $f(x)$ is positive and monotonically decreasing for $x > N$, can be estimated

$$R_N = \sum_{n=N+1}^{\infty} f(n) \leq \int_{N}^{\infty} f(x)\,dx.$$

Example 3.2.4 Consider the series

$$S = \sum_{n=1}^{\infty} \frac{1}{n^4},$$

which is to be computed with three correct decimals. The integral estimate gives

$$R_N \leq \int_N^{\infty} \frac{dx}{x^4} = \frac{1}{3N^3} \leq \frac{1}{2} \cdot 10^{-3},$$

and $N \geq (2000/3)^{1/3} \approx 8.7$. To get three correct decimals, we must then take nine terms in the partial sum. We compute

$$S_9 \approx 1.081937;$$

here we have summed backwards, i.e., started with the smallest terms (cf. Section 2.7). Since our computer has unit roundoff $\mu < 10^{-7}$, we see that the rounding errors in the summation are negligible compared to the other errors. The remainder term estimate then becomes

$$R_9 \leq \frac{1}{3 \cdot 9^3} \leq 0.458 \cdot 10^{-3}.$$

Finally, we get

$$S = 1.081937 \pm 0.458 \cdot 10^{-3} = 1.082 \pm 0.54 \cdot 10^{-3}.$$

If each term in a positive series is less than the corresponding term in another series, whose sum is known (there is an analytic expression for it), then we can estimate the remainder term using the known series.

Comparison with a known series. Assume that

$$0 \leq a_n \leq b_n, \qquad n \geq N+1,$$

and that $T_N = \displaystyle\sum_{n=N+1}^{\infty} b_n$ is known and convergent. Then we have

$$R_N = \sum_{n=N+1}^{\infty} a_n \leq \sum_{n=N+1}^{\infty} b_n = T_N.$$

Often the known series is a geometric series

$$\sum_{n=0}^{\infty} r^n = 1 + r + r^2 + r^3 + \ldots = \frac{1}{1-r}, \qquad |r| < 1.$$

Example 3.2.5   For how large an interval do we get four correct decimals from the approximation

$$e^x \approx 1 + x$$

(the first two terms of the Taylor expansion)?

The remainder term is

$$R = \frac{x^2}{2!} + \frac{x^3}{3!} + \frac{x^4}{4!} + \frac{x^5}{5!} \ldots.$$

First, we see that we cannot easily estimate the remainder by an integral. We rewrite the series so that we can estimate by a geometric series:

$$R = \frac{x^2}{2}(1 + \frac{x}{3} + \frac{x^2}{3 \cdot 4} + \frac{x^3}{3 \cdot 4 \cdot 5} \ldots)$$
$$\leq \frac{x^2}{2}(1 + \frac{x}{3} + \frac{x^2}{3 \cdot 3} + \frac{x^3}{3 \cdot 3 \cdot 3} \ldots) = \frac{x^2/2}{1 - x/3}.$$

By solving the inequality

$$\frac{x^2/2}{1 - x/3} \leq \frac{1}{2} 10^{-4},$$

we find that the approximation $e^x \approx 1 + x$ gives four correct decimals for $0 \leq x \leq 0.00998$.

For this series, we can also estimate the truncation error using Lagrange's remainder term.

3.3   Standard Functions

As we have already noted, it is important to be able to compute the elementary functions very efficiently. In most programming languages for scientific computations, there are standard functions that one can use directly without having to bother with how they have been implemented. However, in some cases, where one works with very simple processors, it may be necessary for the applications programmer to implement the standard functions that are needed.

In this and the following sections, we shall give a brief introduction to some methods for implementing standard functions. First, we give examples of approximations of elementary functions that can be used for the implementations. We do not discuss how such approximations can be derived, since this is outside the scope of this book (some basic principles are mentioned in Chapter 9; we give references at the end of this chapter and Chapter 9, cf. also Chapter 4, where we discuss the implementation of the square root function).

As a rule, the functions are only approximated for small arguments. In the next section, we describe how they can be computed for arbitrary arguments by doing range reduction.

The basic requirements when implementing standard functions are

1) the relative error of the approximation must be smaller than a given tolerance (in general the unit roundoff of the floating point system);
2) the algorithm for computing the approximate function values must be fast.

Since we only approximate the function for small arguments, it is natural to try to use a series expansion around $x = 0$. As an example, consider the approximation by the first four terms of the Maclaurin expansion,

$$\sin x \approx p(x) = x - \frac{x^3}{3!} + \frac{x^5}{5!} - \frac{x^7}{7!}.$$

For $0 \leq x \leq \pi/2$, this approximation has an *absolute error* less than

$$\frac{(\pi/2)^9}{9!} < 1.61 \cdot 10^{-4}$$

(estimate the remainder term for an alternating series).

Maclaurin expansions are accurate for small arguments, but they become bad as we move away from the origin (cf. Example 3.2.5). One can determine the coefficients of a polynomial so that it becomes a good approximation of the function (in our example, the sine function) in *the whole interval* (cf. Chapter 9). The polynomial approximation

$$\sin(\frac{\pi}{2}x) \approx q(x) = x(b_0 + b_1 x^2 + b_2 x^4 + b_3 x^6),$$

where
$$b_0 = 1.570794851,$$
$$b_1 = -0.6459209764,$$
$$b_2 = 0.0794876547,$$
$$b_3 = -0.004362469,$$

has a *relative error* less than 10^{-6} for all x in the interval $[0,1]$. Both this approximation and the Maclaurin expansion above can be computed with five multiplications and three additions: First one computes x^2, and then the polynomial according to the formula

$$q(x) = x(((b_3 x^2 + b_2)x^2 + b_1)x^2 + b_0)$$

(this method is called Horner's rule and will be discussed further in Chapter 4).

Some functions can be computed even more efficiently if they are approximated by a **rational function**

$$\frac{p(x)}{q(x)},$$

where p and q are polynomials. Consider the function $f(x) = 2^x$. If we can approximate this function for small x, then we can compute the usual exponential function from

$$e^x = 2^{x \log_2 e}$$

(in the next section we discuss further the exponential function, and show why it is convenient to use this formula). If we want to approximate 2^x on the interval $[0, \frac{1}{2}]$ by a polynomial

$$2^x \approx p(x),$$

and require the approximation to have a relative error less than 10^{-10}, we must choose $p(x)$ to be a polynomial of degree 6. To evaluate this polynomial, we then need six multiplications and six additions with Horner's rule.

If, instead, we use the approximation

$$2^x \approx \frac{q(x^2) + xs(x^2)}{q(x^2) - xs(x^2)},$$

where

$$q(y) = 20.8189237930062 + y,$$
$$s(y) = 7.2152891511493 + 0.0576900723731y,$$

we get the same accuracy, but the evaluation is considerably faster (provided that division is not much slower than multiplication); only four additions, three multiplications and one division are needed.

It is, of course, no coincidence that a rational function, which is a good approximation of the exponential function, has this form. The following identity holds for the exponential function:

$$2^{-x} = \frac{1}{2^x}.$$

If we approximate 2^x by a rational function $r(x)$ and require that

$$r(-x) = \frac{1}{r(x)},$$

we easily find that it must be possible to write $r(x)$ in the form

$$r(x) = \frac{q(x^2) + xs(x^2)}{q(x^2) - xs(x^2)},$$

where q and s are polynomials.

To conclude, we state that as a rule there are better polynomial approximations of the standard functions than the Maclaurin expansions. For some functions, it is more efficient to approximate by rational functions. In both cases the coefficients are stored, possibly in hardware. Tables of approximations of the standard functions are given in some books; see the references at the end of this chapter.

3.4   Range Reduction

In the preceding section, we gave examples of approximations of standard functions for *small* arguments. In this section, we shall discuss how to do **range reduction**, i.e., use mathematical identities to reduce the evaluation of a function for an arbitrary argument to the evaluation for a small argument.

For trigonometric functions, one of course uses periodicity; we take $\sin x$ as an example. Assume that we have a function, e.g., a polynomial that approximates $\sin x$ well for $0 \le x \le \pi/2$. Using the identity

$$\sin x = \sin(x + 2\pi k),$$

(k is an integer) a given positive argument x can be reduced to the interval $[-\pi, \pi]$:

$$x := x - 2\pi m,$$

for some natural number m. If now $0 \le x \le \pi/2$, then we are ready. Otherwise, if $x > \pi/2$ we can subtract π

$$x := x - \pi.$$

Then x is in the interval $[-\pi/2, 0]$, and we can evaluate $\sin x$ using the identity $\sin(-x) = -\sin x$ and our approximation of the function on the interval $[0, \pi/2]$. We can proceed similarly if, after the first reduction, x is in the interval $[-\pi, -\pi/2]$.

Thus, the total range reduction is

$$x := x - n\pi,$$

where n is an integer. If the given argument is large, however, *cancellation* will occur (see Chapter 2.3). We shall analyze the rounding errors in this computation in a floating point system with unit roundoff μ. We write

$$u = x - n\pi,$$

and assume that x and n are exact, and that π is represented to full accuracy in the floating point system, i.e.,

$$\bar{\pi} = \pi(1 + \epsilon_1),$$

where $|\epsilon_1| \leq \mu$ (in the sequel we assume that all ϵ satisfy such an inequality). The computed approximation $\bar{u} = \text{fl}[x - n\bar{\pi}]$ can then be written

$$\bar{u} = (x - n\bar{\pi}(1 + \epsilon_2))(1 + \epsilon_3) = x(1 + \epsilon_3) - n\pi(1 + \epsilon_1)(1 + \epsilon_2)(1 + \epsilon_3).$$

If we neglect terms that are $O(\mu^2)$, we find that the absolute error can be estimated as

$$|\Delta u| \approx |(x - n\pi)\epsilon_3 - n\pi(\epsilon_1 + \epsilon_2)| \lesssim |u|\mu + 2n\pi\mu.$$

Example 3.4.1   Assume that $x_0 = 1000$, and that we shall compute $\sin x_0$ in IEEE single precision, i.e., in the floating point system $(2, 23, -126, 127)$. We get

$$n = 318, \qquad u = 1000 - 318\pi \approx 0.973536\ldots.$$

The unit roundoff is $\mu = 2^{-24}$, and we have a bound for the absolute error:

$$|\Delta u| \lesssim 2000\mu \leq 1.2 \cdot 10^{-4}.$$

The relative error in u can then be estimated to be

$$\left|\frac{\Delta u}{u}\right| \lesssim \frac{1.2 \cdot 10^{-4}}{0.973} \leq 1.3 \cdot 10^{-4}.$$

The maximal error bound now gives the absolute error in the function value

$$|\Delta(\sin x_0)| \lesssim |\cos u \Delta u| \lesssim \cos 0.973 \cdot 1.2 \cdot 10^{-4} \leq 0.68 \cdot 10^{-4}.$$

Thus, the range reduction gives a relative error in the function value that can be estimated as

$$\left|\frac{\Delta(\sin x_0)}{\sin x_0}\right| \le 0.9 \cdot 10^{-4}.$$

This is, of course, unacceptably large in a floating point system with unit roundoff $\mu \approx 6 \cdot 10^{-8}$.

The cancellation becomes less severe if the range reduction is performed in double precision (or simulated extended precision; see the exercises at the end of the chapter).

Example 3.4.2 The double precision format of the IEEE standard has unit roundoff $\mu = 2^{-53} \approx 1.1 \cdot 10^{-16}$. If the range reduction of the previous example is performed in double precision, we get the estimate

$$\left|\frac{\Delta(\sin x_0)}{\sin x_0}\right| \le 1.6 \cdot 10^{-13}.$$

In the previous section, we indicated that it is suitable to implement the exponential function using the formula

$$e^x = 2^{x \log_2 e}.$$

This is the case if we work in a floating point system with base 2. The computation of $y = e^x$ is performed in four steps:

$$v := x \log_2 e, \quad (\log_2 e \text{ is assumed to be stored accurately})$$
$$z := v - [v], \quad ([v] \text{ denotes the closest integer})$$
$$m := 2^z, \quad (\text{rational approximation})$$
$$y := m \cdot 2^{[v]}.$$

This algorithm is based on the identity

$$2^{x_1+x_2} = 2^{x_1} \cdot 2^{x_2},$$

for the range reduction, so that we only need approximate the exponential function for small arguments.

Example 3.4.3   Assume that we shall implement the exponential function in IEEE single precision, (i.e., we use the floating point system $(2, 23, -126, 127)$, see Section 2.8). Then e^x is to be computed for $0 \leq x \leq 88.72$ (larger arguments give overflow), and the result shall have a relative error less than the unit roundoff μ. We shall examine how the error in the first step of the algorithm is propagated in the computations.

To meet the accuracy requirement, we must execute the statement

$$v := x \log_2 e,$$

in an extended precision format, whose unit roundoff we denote μ_1. Then the computed approximation becomes

$$\bar{v} = v(1 + \epsilon_1)(1 + \epsilon_2), \qquad |\epsilon_i| \leq \mu_1, \quad i = 1, 2,$$

where the last two factors come from the roundoff errors in the floating point representation of $\log_2 e$ and the multiplication, respectively. Provided that $[\bar{v}] = [v]$ (in the sequel, we assume that this is the case), we get

$$\bar{z} = \bar{v} - [\bar{v}] \approx v + v(\epsilon_1 + \epsilon_2) - [v] = z + v(\epsilon_1 + \epsilon_2).$$

The absolute error in $\bar{z}$ is then bounded by

$$|\Delta z| \leq 2v\mu_1.$$

The error in $\bar{z}$ is propagated in the computation of the significand m. We can investigate this using the maximal error estimate

$$\Delta m \approx \frac{dm}{dz} \Delta z = m \log 2 \, \Delta z,$$

and, for the relative error, we have the bound

$$\left| \frac{\Delta m}{m} \right| \lesssim \log 2 \, |\Delta z| \lesssim 2 \log 2 \, v\mu_1.$$

But $v = x \log_2 e = x/\log 2$, so that

$$\left| \frac{\Delta m}{m} \right| \lesssim 2x\mu_1.$$

We want the relative error in m to be smaller than μ, and thus we require that

$$\left| \frac{\Delta m}{m} \right| \leq 2^{-24}.$$

How many extra bits are then needed for the significand of the extended format?

The relative error $|\Delta m/m|$ shall be less than 2^{-24} for all $x \in [0, 88.72]$. We get

$$2x\mu_1 \leq 2 \cdot 88.72\mu_1 \leq 2^{-24}.$$

The single precision significand has 24 bits. Assume that the significand of the extended format has s extra bits. The inequality then becomes

$$2 \cdot 88.72 \cdot 2^{-24-s} \leq 2^{-24},$$

which gives

$$s \geq \frac{\log(2 \cdot 88.72)}{\log 2} \approx 7.47.$$

Thus, the significand of the extended format must have at least eight extra bits. Note that the extended single precision format in the IEEE standard has at least eight extra bits, see Section 2.8.

3.5 An Algorithm for the Evaluation of Trigonometric Functions

Earlier in this chapter, we discussed how elementary functions can be evaluated using polynomial and rational approximations. These are the classical techniques that are relatively easy to implement, e.g., in assembler code. When a processor is designed using VLSI technology, however, it may be advantageous to implement the elementary functions at a lower level, i.e., with simpler operations than multiplications. This is possible because, using VLSI, one can construct more complicated and larger (meaning with more simple components) systems in hardware than with earlier technology. What is gained is speed; function values can be computed in approximately the same time as it takes to perform one or a couple of divisions.

In this section, we shall describe a method for computing the trigonometric functions, which is called **CORDIC** (COordinate Rotation DIgital Computer). It was first presented in 1959, and it has been used for INTEL's 8087 processor. It can also be generalized to compute square roots, exponential functions, and the logarithm function. Further, multiplication and division can be implemented using this technique. We want to emphasize that by necessity our presentation is simplified. The aim of this section is not to describe an actual implementation, but rather to discuss some of the principles that form the basis of real implementations.

Let $\beta > 0$ be an angle given in radians, and assume that we shall compute $\sin\beta$. The computer is assumed to have a binary floating point system

$(2, t, L, U)$. The accuracy requirement for $\sin x$ is that the relative error should be smaller than the unit roundoff μ. First, we assume that the angle β has been reduced to the interval $[0, \pi/2]$. Second, we assume that the computations in the Cordic algorithm are executed in a fixed point format with $2t$ bits. This means that an angle given in the floating point system must be transformed to the fixed point format. Since $0 \le \beta \le \pi/2$, we can write β in the floating point system in the form

$$\beta = m2^{-e},$$

where m is the normalized significand, and e is a nonnegative integer. To transform the floating point number β to fixed point format, we shift the normalized significand m in a register e steps to the right. If e is large, then m will be shifted partly or completely outside the register, and loss of accuracy will occur. Therefore, we cannot use the Cordic algorithm for small angles, at least not as it is described here. We thus start by discussing how to approximate $\sin \beta$ for small β.

We know that

$$\sin \beta \approx \beta$$

is a good approximation for small angles. We shall now determine the largest angle for which it can be used. The relative truncation error is approximately (cf. Section 3.2)

$$\frac{\beta^3/6}{\beta} = \frac{\beta^2}{6}.$$

For which β is this quantity less than the unit roundoff? With $\beta = m2^{-e}$, we get the inequality

$$\frac{m^2 2^{-2e}}{6} \le \mu = \frac{1}{2} 2^{-t}.$$

Since the inequality must hold for all m in the interval $[1, 2)$, we get

$$\frac{4}{6} 2^{-2e} \le \frac{1}{2} 2^{-t}.$$

We can write the inequality as

$$e \ge \frac{1}{2}(-\log_2 0.75 + t);$$

this means that the relative error of the approximation $\sin \beta \approx \beta$ is less than μ for all $\beta = m2^{-e}$, where $e \ge (t+1)/2$. Therefore, we need only use the Cordic algorithm for angles

$$\beta = m2^{-e}, \qquad 0 \le e \le \lfloor (t+1)/2 \rfloor,$$

where $\lfloor p \rfloor$ denotes the largest integer smaller than or equal to p.

Next, consider the conversion from floating point format to fixed point format. Let the normalized significand be $m = 1.d_1 d_2 \ldots d_t$, where the d_i are binary digits. We can now illustrate the conversion in the register with $2t$ bits symbolically as

$$1.d_1 d_2 \ldots d_t 00 \ldots 0 \quad \longrightarrow \quad 0.00 \ldots 01 d_1 d_2 \ldots d_t 0 \ldots 0,$$

where the number of zeroes that are introduced to the left is equal to e. Since we have $e \leq \lfloor (t+1)/2 \rfloor$, all the digits in m will be inside the register after the shift. Hence, no errors are introduced in the conversion from floating point to fixed point format.

We now describe the basic idea behind the Cordic algorithm without making any assumptions about how the angle β is represented. β is given in radians, and we shall compute $\sin \beta$. Geometrically, this means that we shall determine the y-coordinate for the vector v in Figure 3.5.1.

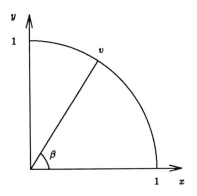

Figure 3.5.1

This can be done as follows: We start with the vector

$$v_0 = \begin{pmatrix} 1 \\ 0 \end{pmatrix}.$$

Rotate v_0 by a given angle γ_0 (in the positive direction), and denote the corresponding vector v_1.

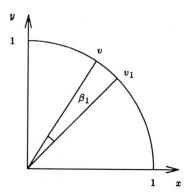

Figure 3.5.2

Define $\beta_0 = \beta$, and put

$$\beta_1 = \beta_0 - \gamma_0.$$

If $\beta_1 > 0$, then rotate v_1 the angle γ_1 in positive direction, denote the corresponding vector v_2 and put

$$\beta_2 = \beta_1 - \gamma_1.$$

(If $\beta_1 < 0$, i.e., if v_1 has passed v, rotate v_1 the angle $-\gamma_1$ instead.) In Figure 3.5.3 the new vector v_2 has passed v, and the angle β_2 is negative.

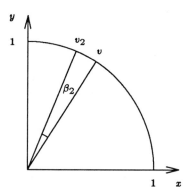

Figure 3.5.3

In this way, we can go on and successively rotate the vector by angles $\gamma_0, \gamma_1, \gamma_2, \ldots$, in the positive or negative direction depending on if we have passed v or not. The procedure stops when the vector v_i approximates v to the required accuracy. Note that so far we have not decided which sequence $\gamma_0, \gamma_1, \ldots$ we are going to use.

To make the description more precise, we define a **rotation matrix**

$$P = \begin{pmatrix} \cos\gamma & -\sin\gamma \\ \sin\gamma & \cos\gamma \end{pmatrix}.$$

The matrix P is **orthogonal**, i.e.,

$$P^{-1} = P^T = \begin{pmatrix} \cos\gamma & \sin\gamma \\ -\sin\gamma & \cos\gamma \end{pmatrix}.$$

If we multiply a vector by a rotation matrix, the length remains the same. Let

$$v_1 = \begin{pmatrix} x_1 \\ y_1 \end{pmatrix} = P_0 v_0 = \begin{pmatrix} \cos\gamma_0 & -\sin\gamma_0 \\ \sin\gamma_0 & \cos\gamma_0 \end{pmatrix} \begin{pmatrix} x_0 \\ y_0 \end{pmatrix};$$

the length of the vector v_1 is given by

$$(x_1^2 + y_1^2)^{1/2} = ((x_0\cos\gamma_0 - y_0\sin\gamma_0)^2 + (x_0\sin\gamma_0 + y_0\cos\gamma_0)^2)^{1/2}$$
$$= (x_0^2 + y_0^2)^{1/2}.$$

The rotation of a vector v_0 by an angle γ_0 is equivalent to multiplying $v_0 = (x_0, y_0)^T$ by the matrix P_0:

$$v_1 = P_0 v_0 = \begin{pmatrix} \cos\gamma_0 & -\sin\gamma_0 \\ \sin\gamma_0 & \cos\gamma_0 \end{pmatrix} \begin{pmatrix} x_0 \\ y_0 \end{pmatrix}.$$

This is seen immediately if we choose $v_0 = (1,0)^T$.

The algorithm can now be described:

Cordic algorithm, preliminary version: Let $\gamma_0, \gamma_1, \ldots$ be a given sequence of rotation angles, and let

$$v_0 = \begin{pmatrix} 1 \\ 0 \end{pmatrix}, \qquad \beta_0 = \beta,$$

be initial values. The vectors v_i and the angles β_i are computed recursively from

$$v_{i+1} = P_i v_i, \qquad \beta_{i+1} = \beta_i - \sigma_i \gamma_i, \qquad i = 0, 1, 2, \ldots,$$

where
$$\sigma_i = \text{sign}(\beta_i),$$
$$P_i = \begin{pmatrix} \cos\gamma_i & -\sigma_i \sin\gamma_i \\ \sigma_i \sin\gamma_i & \cos\gamma_i \end{pmatrix}.$$

This is an iterative method for computing the coordinates for the vector v , i.e., $\cos\beta$ and $\sin\beta$, and it can be shown to converge for a suitable choice of angles $\{\gamma_0, \gamma_1, \ldots\}$ (if, for instance, we choose $\gamma_i = 2^{-i}\pi/4$, then the algorithm reminds us of the bisection method).

In the preliminary version, the algorithm is not particularly fast or simple, since in each step of the recursion we must perform four multiplications and two additions. We shall now see that the algorithm becomes a finite recursion when it is applied to an angle β given in fixed point format, and that it can be implemented efficiently in a computer, if the angles γ_i are chosen appropriately. We again remind the reader that the presentation is greatly simplified.

Consider the matrix-vector multiplication

$$P_i v_i = \begin{pmatrix} c_i & -\sigma_i s_i \\ \sigma_i s_i & c_i \end{pmatrix} v_i,$$

where we have put $c_i = \cos\gamma_i$ and $s_i = \sin\gamma_i$. Now rewrite

$$P_i v_i = c_i Q_i v_i = c_i \begin{pmatrix} 1 & -\sigma_i t_i \\ \sigma_i t_i & 1 \end{pmatrix} v_i,$$

where

$$t_i = \frac{s_i}{c_i} = \tan\gamma_i.$$

If we choose the angles so that

$$t_i = 2^{-i},$$

then the multiplication by the matrix Q_i becomes very simple — two shift operations and two additions:

$$Q_i v_i = \begin{pmatrix} 1 & -\sigma_i 2^{-i} \\ \sigma_i 2^{-i} & 1 \end{pmatrix} \begin{pmatrix} x_i \\ y_i \end{pmatrix} = \begin{pmatrix} x_i - \sigma_i 2^{-i} y_i \\ \sigma_i 2^{-i} x_i + y_i \end{pmatrix}.$$

Assume that also x_i and y_i are stored in fixed point format. Now, it is obvious that the algorithm is finite; after $2t$ steps, the shift is so large that everything is shifted outside the register, and the matrix Q becomes an identity matrix in the fixed point arithmetic.

The recursion for β_i is

$$\beta_{i+1} = \beta_i - \sigma_i \gamma_i, \quad \gamma_i = \arctan 2^{-i}, \quad i = 0, 1, 2, \ldots.$$

Since $\arctan x \le x$, we see that this recursion is finite with $2t$ steps; when $i = 2t$, then γ_i can no longer be represented in the fixed point format.

The approximation of the vector v that we want to compute is

$$v_{2t} = c_{2t-1}c_{2t-2}\cdots c_1 c_0 Q_{2t-1} Q_{2t-2} \cdots Q_1 Q_0 v_0.$$

Now put

$$\tau = c_{2t-1}c_{2t-2}\cdots c_1 c_0.$$

Note that τ can be computed, once and for all, since the angles γ_i are given at the outset. Define

$$\tilde{v}_0 = \tau v_0 = \begin{pmatrix} \tau \\ 0 \end{pmatrix}.$$

v_{2t} is now given by

$$v_{2t} = Q_{2t-1} Q_{2t-2} \cdots Q_1 Q_0 \tilde{v}_0.$$

We summarize:

THE CORDIC ALGORITHM:

$$v := \begin{pmatrix} \tau \\ 0 \end{pmatrix};$$

for $i := 0, 1, 2, \ldots, 2t - 1$ **do**
$\qquad \sigma := \text{sign}(\beta);$

$$v := \begin{pmatrix} 1 & -\sigma 2^{-i} \\ \sigma 2^{-i} & 1 \end{pmatrix} v;$$

$\qquad \beta := \beta - \sigma \gamma_i;$

Now $v = \begin{pmatrix} \cos \beta \\ \sin \beta \end{pmatrix}$ in the fixed point format.

We make the following observations:

1. The algorithm uses fixed point arithmetic only, which is faster than floating point arithmetic. The operations are simple — shifts and additions. The only logical operation is a test of the sign of β.
2. The angles $\gamma_i = \arctan 2^{-i}, i = 0, 1, 2, \ldots, 2t - 1$, must be stored. But, for small angles, we have $\arctan \gamma_i = \gamma_i$ in the finite precision, and, therefore, we need not store these angles. This way we can reduce the size of the table.

Exercises

1. Use series expansions to compute
   a) $\sin 0.1$,
   b) $e^{0.12}$
   with three correct decimals.

2. Let $f(x)$ be a positive, monotonically decreasing function for large x. Draw two figures that illustrate the inequalities

$$f(n+1) + f(n+2) + \ldots > \int_{n+1}^{\infty} f(x)\,dx,$$

$$f(n+1) + f(n+2) + \ldots < \int_{n}^{\infty} f(x)\,dx.$$

Use the inequalities to compute

$$S = \sum_{n=1}^{\infty} \frac{1}{n^3}$$

with two correct decimals.

3. Compute

$$S = \sum_{n=0}^{\infty} \frac{(-1)^n}{n^6 + 1}$$

with three correct decimals.

4. Assume that we shall compute $\sin x$ in IEEE single precision, and consider the problem of doing range reduction,

$$u = x_0 - n\pi.$$

In order to avoid cancellation without having to use double precision, we can simulate extended precision by writing π in the form

$$\pi = \pi_0 + r,$$

where π_0 is exactly representable in the floating point system, and where r is small. The reduced argument u is now computed,

$$u = (x_0 - n\pi_0) - nr.$$

Since π_0 and n are represented exactly in the floating point system, the first computation can be performed without rounding errors (if n is not extremely large). Let

$$x_0 = 1000,$$
$$\pi_0 = 3.140625 = (101.001001)_2,$$
$$r = 0.000967654\ldots.$$

Assume that r is represented to full accuracy in single precision. Estimate the relative error of the computed approximation $\bar{u}$. (See Examples 3.4.1 and 3.4.2).

5. The range reduction for a trigonometric function is, of course, the most difficult when the argument is close to a multiple of π, since in this case the cancellation can be disastrous. Let $u = x_0 - n\pi$, and assume that u is small. Show that

$$\left| \frac{\Delta(\sin x_0)}{\sin x_0} \right| \lesssim \left| \frac{\Delta u}{u} \right|.$$

Assume that the range reduction is done in IEEE double precision. For how large an interval around $x = \pi$ can we *not* compute $\sin x$ to full accuracy (i.e., with an error smaller than the unit roundoff) in single precision? Only the errors from the range reduction are to be taken into account.

6. Show that the recursions in the Cordic algorithm for the two components of the v vector are independent for $i > t$, and that

$$\begin{pmatrix} x_{i+1} \\ y_{i+1} \end{pmatrix} = \begin{pmatrix} x_i + \sigma_i 2^{-i} y_t \\ y_i - \sigma_i 2^{-i} x_t \end{pmatrix}, \qquad i = t+1, t+2, \ldots, 2t-1.$$

7. Compute τ in the Cordic algorithm for $t = 23$.

8. For which values of i do we have $\arctan 2^{-i} = 2^{-i}$ in a fixed point arithmetic with 46 binary digits and one integer digit?

References

Much of the theory for remainder term estimates can be found in textbooks in analysis. More about transformations of series can be found in

C.-E. Fröberg, *Numerical Mathematics, Theory and Computer Applications*, The Benjamin/Cummings Publishing Company, Menlo Park, 1985.

In the following books there are tables of polynomial and rational approximations of standard functions.

J. F. Hart et al., *Computer Approximations*, John Wiley and Sons, New York, 1968.

W. J. Cody, Jr. and W. Waite, *Software Manual for the Elementary Functions*, Prentice–Hall, Englewood Cliffs, New Jersey, 1980.

Algorithms for standard functions have been developed that circumvent the problems related to range reduction. Thus, the functions are correct to the last digit for very large intervals of arguments. This work is described in

R. C. Agarwal et al., *New scalar and vector elementary functions for the IBM System/370*, IBM J. Res. Develop. **30**(1986), 126–143.

4   Nonlinear Equations

4.1   Introduction

In this chapter, we shall study methods for solving a nonlinear equation

$$f(x) = 0,$$

where f is assumed to be a continuous, real valued function of one real variable. Whenever appropriate, we shall assume that f is differentiable. The root x^* that we are looking for is assumed to be **simple**. In Section 4.7, we shall briefly discuss the solution of systems of nonlinear equations.

In general, a nonlinear equation $f(x) = 0$ cannot be solved analytically, i.e., the solution cannot be given explicitly in a formula consisting of elementary algebraic operations and elementary functions. Algebraic equations (equations where $f(x)$ is a polynomial) of degree less than or equal to four can be solved analytically, but it can be proved not to be possible in general for algebraic equations of higher degree. Nor can transcendental equations (equations involving transcendental functions like the exponential function and trigonometric functions) be solved analytically in general. In some cases, the function $f(x)$ is not even known explicitly; e.g., it can be defined via the solution of a differential equation (see Chapter 9). In such cases, it is necessary to use numerical methods for solving the equation.

The basic idea that is used in most numerical methods for solving equations is first to determine an initial approximation x_0 of the root. Starting from x_0, a sequence $(x_n)_{n=0}^{\infty}$ is constructed that converges to the root,

$$\lim_{n \to \infty} x_n = x^*.$$

A method that produces such a sequence is called an **iteration method**.

We shall first describe some ways of constructing iteration methods, and then discuss convergence criteria. Often we want the sequence to converge rapidly to the root. We define order of convergence, and examine a few methods from this point of view. In practice, we can only compute finitely many numbers in the sequence. We discuss error estimation, and the accuracy that can be obtained in finite precision arithmetic. We also describe a method for implementing the square root function in a computer.

4.2   Some Iteration Methods

We shall discuss here mainly how to solve equations using simple computing tools like a pocket calculator. The methodology is essentially the same when a computer is used.

Throughout this section, we shall use the equation

$$f(x) = x - e^{-x} = 0$$

as an illustration. When solving an equation numerically, we must first obtain an initial approximation x_0. This can be done as follows:

    a. graphing the function,
    b. tabulating the function,
    c. using the bisection method.

The root of our illustration equation is the x coordinate of the point where the curves $y = x$ and $y = e^{-x}$ intersect. From Figure 4.2.1, we see that the root is close to 0.6.

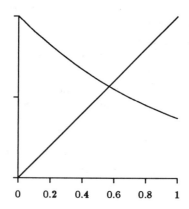

$$0 \quad 0.2 \quad 0.4 \quad 0.6 \quad 0.8 \quad 1$$

Figure 4.2.1   The curves $y = x$ and $y = e^{-x}$.

Using graphs, we often get more information than just the approximate position of the root. E.g., from Figure 4.2.1 we see that, since the function $g(x) = x$ is monotonically increasing and $h(x) = e^{-x}$ is monotonically decreasing, the curves intersect only once, i.e., the equation $f(x) = x - e^{-x} = 0$ has exactly one root.

We tabulate the function $f(x)$ in the neighborhood of 0.6.

x	e^{-x}	$f(x)$
0.5	0.61	−0.11
0.6	0.55	0.05

Table 4.2.2.   The function $f(x) = x - e^{-x}$.

From Table 4.2.2, we see that the root is between 0.5 and 0.6.

In order to systematically construct better approximations, we can use the **bisection method**. This method is based on the idea of successively enclosing the root in smaller and smaller intervals. In our example, we start with the interval $(0.5, 0.6)$.

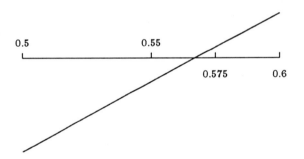

Figure 4.2.3   The bisection method.

Since $f(0.55)$ is negative, the root is to the right of 0.55 (see Figure 4.2.3), and the next interval is $(0.55, 0.6)$. The midpoint 0.575 of this interval is to the right of the root, and hence the next interval is $(0.55, 0.575)$, etc.

We immediately see that this method always converges, but the rate of convergence is slow. After n steps, the root is enclosed in an interval with the length 2^{-n} times the length of the original interval. If, in our example, we start with the interval $(0.5, 0.6)$, we must halve the interval 17 times to get an interval shorter than 10^{-6} (the midpoint of that interval is an approximation to the root with six correct decimals).

Because of the slow convergence, the bisection method should only be used to obtain a starting approximation for a method with faster convergence. Such methods are based on the following principle: Using information about the function f and about the approximate position of the root, compute a new and more accurate approximation. The more information about the function we use, the faster convergence we get.

Let us for a moment return to the bisection method, and see what information is used there. When starting from an interval, we compute an approximate function value at the midpoint, but we do not use most of that information. It is only the sign of the function that determines which interval half to select. Assume that we have the situation illustrated in Figure 4.2.4: a large positive function value and a small negative function value.

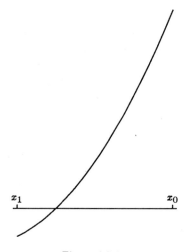

Figure 4.2.4

The small function value $f(x_1)$ indicates that the root is closer to x_1 than x_0. If we use the information contained in the magnitude of the function values, we should get faster convergence than in the bisection method. This can be done as follows: Approximate the function by a straight line, and take its point of intersection with the x-axis as the new approximation of the root, see Figure 4.2.5.

We shall now derive an expression for x_2. The equation of the straight through the points $(x_0, f(x_0))$ and $(x_1, f(x_1))$ is

$$y - f(x_1) = \frac{f(x_1) - f(x_0)}{x_1 - x_0}(x - x_1).$$

We get its intersection with the x-axis by putting $y = 0$ and solving for x

$$x_2 = x_1 - f(x_1)\frac{x_1 - x_0}{f(x_1) - f(x_0)}.$$

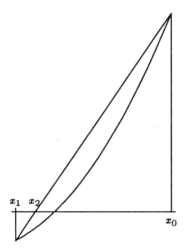

Figure 4.2.5 Approximation by a straight line.

We can now continue in a way that is similar to the bisection method: Enclose the root in intervals, where at the endpoints there is one positive and one negative function value. This variant is called **Regula Falsi** and is illustrated in Figure 4.2.6.

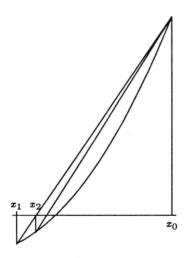

Figure 4.2.6 Regula Falsi.

This method is robust (safe), since the root will be enclosed in smaller and smaller intervals. It has the same weakness as the bisection method, however: it converges relatively slowly.

Another variant is to apply the iteration formula

$$x_{n+1} = x_n - f(x_n) \frac{x_n - x_{n-1}}{f(x_n) - f(x_{n-1})}, \qquad n = 1, 2, \ldots,$$

as it stands (we get the formula from the derivation above). Thus, we do not require the function f to have different signs for x_{n-1} and x_n. This is the **secant method** (the straight line that approximates the function is called a secant).

The secant method

$$x_{n+1} = x_n - f(x_n) \frac{x_n - x_{n-1}}{f(x_n) - f(x_{n-1})}, \qquad n = 1, 2, \ldots.$$

The secant method is illustrated in Figure 4.2.7.

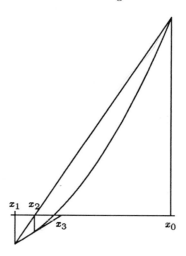

Figure 4.2.7   The secant method.

As is seen in Figure 4.2.7, it can happen that both approximations in the secant method are on the same side of the root. Therefore, it is not as robust as Regula Falsi; if the initial approximations are not close enough to the root, it may happen that the method gives a sequence that does not converge to the desired root (see Figure 4.2.8).

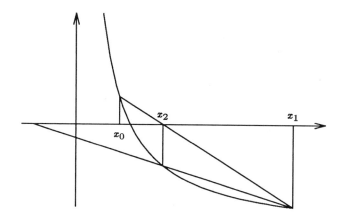

Figure 4.2.8   Example of divergence with the secant method.

If the initial approximations for the secant method are chosen close to the root, and the function is well-behaved, e.g., as in Figure 4.2.7, then the secant method converges faster than Regula Falsi. In Table 4.2.9 we give the results for the secant method applied to the equation $f(x) = x - e^{-x} = 0$, with the initial interval $(0.55, 0.575)$ (the interval obtained with the bisection method at the beginning of this section).

n	x_n	$f(x_n)$
0	0.55	-0.026950
1	0.575	0.012295
2	0.567168	0.000038
3	0.567143	$5.28 \cdot 10^{-8}$

Table 4.2.9   The secant method applied to the equation $f(x) = x - e^{-x} = 0$, which has the root $x^* = 0.567143 \pm 0.5 \cdot 10^{-6}$.

An alternative straight line approximation to the function $f(x)$ is to draw a tangent to the curve $y = f(x)$. Assume that we have an approximation x_0 of the root; we draw the tangent to the curve at the point $(x_0, f(x_0))$ and let the intersection of the tangent and the x-axis be the next approximation

x_1 of the root (see Figure 4.2.10).

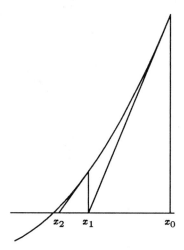

Figure 4.2.10

The equation of the tangent is

$$y - f(x_0) = f'(x_0)(x - x_0).$$

We get its intersection with the x-axis by putting $y = 0$ and solving for x:

$$x_1 = x_0 - \frac{f(x_0)}{f'(x_0)}.$$

We now compute a new approximation of the root starting from x_1, and so on. The general formula is:

Newton–Raphson's method

$$x_{n+1} = x_n - \frac{f(x_n)}{f'(x_n)}, \qquad n = 0, 1, 2, \ldots.$$

Here we have made a geometric derivation of Newton–Raphson's method. It can also be derived analytically. Given an approximation x_n of the root, we can formulate the problem of solving the equation $f(x) = 0$ as follows: Find h such that $f(x_n + h) = 0$. We now approximate the function using the first two terms of the Taylor expansion around x_n:

$$f(x_n + h) \approx f(x_n) + h f'(x_n).$$

If $f'(x_n) \neq 0$, we can put the right hand side equal to zero and solve for h. Since we do not get the exact value of h, we use an index n:

$$h_n = -\frac{f(x_n)}{f'(x_n)}.$$

This is a correction of our approximation of the root, and we denote the new approximation x_{n+1}:

$$x_{n+1} = x_n + h_n = x_n - \frac{f(x_n)}{f'(x_n)}.$$

In Newton–Raphson's method we use both the function value and the derivative when we compute the new approximation of the root. Therefore, we can expect the method to give a rapidly converging sequence. As an example, we solve our illustration equation. Here, we have

$$f(x) = x - e^{-x}, \qquad f'(x) = 1 + e^{-x};$$

The results obtained with $x_0 = 0.55$ are given in Table 4.2.11.

n	x_n	$f(x_n)$	$f'(x_n)$	$f(x_n)/f'(x_n)$
0	0.55	−0.026950	1.576950	−0.017090
1	0.567090	−0.000084	1.567174	−0.000053
2	0.567143	$-4.55 \cdot 10^{-7}$		

Table 4.2.11   The equation $f(x) = x - e^{-x} = 0$ solved using Newton–Raphson's method.

In the next section we shall study in some more detail convergence criteria and rate of convergence of iteration methods. For that discussion, it is convenient to write Newton–Raphson's method in the following form

$$x_{n+1} = \varphi(x_n),$$

where the **iteration function** $\varphi(x)$ is

$$\varphi(x) = x - \frac{f(x)}{f'(x)}.$$

We also see that the equation $x = \varphi(x)$ is an equivalent way of writing $f(x) = 0$ (the equation $f(x) = 0$ can be put into the form $x = \varphi(x)$ by elementary mathematical operations, and vice versa). Since x^* is a root of the equation $f(x) = 0$, we have

$$x^* = \varphi(x^*).$$

x^* is called a **fixed point** of the map $x \to \varphi(x)$, and an iteration method $x_{n+1} = \varphi(x_n)$ is called a **fixed point iteration**.

Fixed point iterations can be obtained by making other equivalent reformulations than the one that leads to Newton–Raphson's method. For instance, our illustration equation can be written

$$x = e^{-x},$$

and the corresponding iteration method is

$$x_{n+1} = e^{-x_n}.$$

This method converges to the root (see Table 4.2.12), albeit slowly.

n	x_n
0	0.55
1	0.57695
2	0.561609
3	0.570291
4	0.565361
⋮	
16	0.567141
17	0.567144
18	0.567143

Table 4.2.12   The fixed point iteration $x_{n+1} = e^{-x_n}$.

Another equivalent reformulation is

$$x = -\log x;$$

the iteration

$$x_{n+1} = -\log x_n$$

is divergent, however (Table 4.2.13).

n	x_n
0	0.55
1	0.597837
2	0.514437
3	0.664682
4	0.408447
5	0.895394
6	0.110492
7	2.202816
8	−0.789737

Table 4.2.13   The divergent fixed point iteration $x_{n+1} = -\log x_n$.

Obviously, we must analyze criteria for the function $\varphi(x)$ to give a convergent fixed point iteration $x_{n+1} = \varphi(x_n)$. This is done in the next section.

4.3   Convergence Criteria and Rate of Convergence

Consider the fixed point iteration

$$x_{n+1} = \varphi(x_n).$$

Defining

$$\epsilon_n = x_n - x^*,$$

we see that the convergence of the sequence $(x_n)_{n=0}^{\infty}$ to x^* is equivalent to the convergence of the sequence $(\epsilon_n)_{n=0}^{\infty}$ to 0.

We can now write

$$\epsilon_n = x_n - x^* = \varphi(x_{n-1}) - \varphi(x^*)$$

(since $x^* = \varphi(x^*)$). Using the mean value theorem, we get

$$\epsilon_n = x_n - x^* = \varphi'(\xi_n)(x_{n-1} - x^*) = \varphi'(\xi_n)\epsilon_{n-1}.$$

If $|\varphi'(\xi_n)| \leq m < 1$ for some constant m, we have

$$|\epsilon_n| \leq m\,|\epsilon_{n-1}| \leq |\epsilon_{n-1}|.$$

The condition $|\varphi'(x)| \leq m < 1$ close to x^* is a sufficient condition for convergence, since then

$$|\epsilon_n| \leq m\,|\epsilon_{n-1}| \leq m^2\,|\epsilon_{n-2}| \leq \ldots \leq m^n\,|\epsilon_0|,$$

and $m^n \to 0$ as $n \to \infty$.

Theorem 4.3.1   Assume that the iteration function $\varphi(x)$ has a real fixed point x^*, and that

$$|\varphi'(x)| \leq m < 1,$$

for all $x \in J$, where J is an interval around x^*,

$$J = \{x \mid |x - x^*| \leq \delta\},$$

for some δ. If $x_0 \in J$, then we have

a) $x_n \in J$, $n = 1, 2, 3, \ldots$,

b) $\lim_{n \to \infty} x_n = x^*$,

c) x^* is the only root of the equation $x = \varphi(x)$ in J.

Proof.   a) is proved by induction. Assume that $x_{n-1} \in J$. The mean value theorem gives

$$x_n - x^* = \varphi(x_{n-1}) - \varphi(x^*) = \varphi'(\xi_n)(x_{n-1} - x^*),$$

and, since ξ_n lies between x_{n-1} and x^*, ξ_n must lie in J. We get

$$|x_n - x^*| \le m\,|x_{n-1} - x^*| \le m\delta < \delta,$$

which means that $x_n \in J$, and we have proved a).

From the above argument, we see that

$$|x_i - x^*| = |\varphi'(\xi_i)|\,|x_{i-1} - x^*| \le m\,|x_{i-1} - x^*|\,,$$

and, since all ξ_i lie in J, we get

$$|x_n - x^*| \le m\,|x_{n-1} - x^*| \le \ldots \le m^n\,|x_0 - x^*|\,.$$

From $m < 1$, we have

$$\lim_{n \to \infty} |x_n - x^*| = 0,$$

and we have shown b).

Uniqueness is proved by contradiction. Assume that $x^{**} = \varphi(x^{**})$, $x^* \ne x^{**}$ and $x^{**} \in J$. The mean value theorem and the assumptions give

$$|x^* - x^{**}| = |\varphi'(\xi)|\,|x^* - x^{**}| \le m\,|x^* - x^{**}| < |x^* - x^{**}|\,,$$

which is a contradiction.   ∎

Example 4.3.2   We examine the two fixed point iterations

$$x_{n+1} = e^{-x_n}, \qquad x_{n+1} = -\log x_n,$$

for the equation $f(x) = x - e^{-x} = 0$. In the first case, we have

$$\varphi(x) = e^{-x}, \qquad \varphi'(x) = -e^{-x}$$

and

$$|\varphi'(x)| \approx 0.567 < 0.6,$$

close to the root $x^* = 0.567143 \pm 0.5 \cdot 10^{-6}$. According to the theory, the iteration is convergent, which we also found experimentally in the previous section.

In the second case, we have

$$\varphi(x) = -\log x, \qquad \varphi'(x) = \frac{-1}{x},$$

and

$$|\varphi'(x)| \approx \frac{1}{0.567} > 1,$$

close to the root. Hence the iteration is divergent.

From the proof of Theorem 4.3.1, we see that the smaller m is, the faster the iteration converges. E.g., for the first method in Example 4.3.2, we have $m \approx 0.6$. Therefore, this method should have slower convergence than the bisection method, which we can also see from the results in the previous section. Let us now analyze the rate of convergence of Newton–Raphson's method. There we have

$$\varphi(x) = x - \frac{f(x)}{f'(x)},$$

and

$$\varphi'(x) = \frac{f(x)f''(x)}{(f'(x))^2}.$$

We see that $\varphi'(x^*) = 0$. Therefore Newton–Raphson's method should converge very rapidly once you are close to the root. Let us examine this method in some more detail. As before, we have

$$x_{n+1} - x^* = \varphi(x_n) - \varphi(x^*),$$

where $\varphi(x) = x - f(x)/f'(x)$. We now use the Taylor expansion of $\varphi(x_n)$ around x^*:

$$\varphi(x_n) = \varphi(x^*) + (x_n - x^*)\varphi'(x^*) + \frac{(x_n - x^*)^2}{2}\varphi''(\eta),$$

where η lies between x_n and x^*. Since $\varphi'(x^*) = 0$, we get

$$x_{n+1} - x^* = \frac{(x_n - x^*)^2}{2}\varphi''(\eta).$$

This shows that Newton–Raphson's method converges faster than fixed point iterations in general. In order to compare the rate of convergence of different iteration methods, we make the following definition.

Let $x_0, x_1, x_2, \ldots$ be a sequence converging to x^*, and put $\epsilon_n = x_n - x^*$. The **order of convergence** of the sequence is said to be p, if p is the largest positive number such that

$$\lim_{n \to \infty} \frac{|\epsilon_{n+1}|}{|\epsilon_n|^p} = C < \infty.$$

C is called the **asymptotic error constant**. For $p = 1$ and $p = 2$, the convergence is called **linear** and **quadratic**, respectively.

Often we say that *an iteration method has order of convergence p*, if it generates sequences with this order of convergence. In general, fixed point iterations have linear convergence with asymptotic error constant $\varphi'(x^*)$. Newton–Raphson's method has quadratic convergence, and its asymptotic error constant is

$$\frac{\varphi''(x^*)}{2} = \frac{f''(x^*)}{2f'(x^*)}.$$

The quadratic convergence can be observed in the following example.

Example 4.3.3   For $a > 0$, we can compute $\sqrt{a}$ by solving the equation $x^2 - a = 0$. Newton–Raphson's method applied to the equation gives

$$x_{n+1} = \frac{1}{2}\left(x_n + \frac{a}{x_n}\right).$$

We choose $a = 3$ and $x_0 = 2$.

n	x_n	$x_n - \sqrt{3}$
0	2	$2.7 \cdot 10^{-1}$
1	1.75	$1.8 \cdot 10^{-2}$
2	1.7321	$4.9 \cdot 10^{-5}$
3	1.7320508	10^{-8}

We return to the computation of square roots in Section 4.6.

The following rule of thumb can be used for Newton–Raphson's method (note, however, that it is a *rule of thumb*, not a general mathematical truth): "*the number of correct decimals are doubled in every iteration.*"
   The following can be shown:

> Newton–Raphson's method always converges (to a simple root) if the initial approximation x_0 is chosen close enough to the root x^*.

If a small interval containing the root x^* is known, then it is often easy to verify that the convergence condition for Newton–Raphson's method,

$$|\varphi'(x)| = \left|\frac{f(x)f''(x)}{(f'(x))^2}\right| \leq m < 1,$$

is satisfied. In most cases, one does not bother to check convergence in advance, as a divergent iteration will show up very quickly. There are some important special cases, where it its quite easy to prove that Newton–Raphson's method converges for arbitrary initial approximations, see Section 4.6. The general theory for Newton–Raphson's method can be found in the references given at the end of this chapter.

4.4 Error Estimation

When approximations of a root x^* are computed using an iteration method, there will be rounding errors. It can be seen from Theorem 4.3.1 that, as long as the errors are small (so that all approximations are inside the interval J), these errors do not influence the final accuracy. In this respect iteration methods are *self-correcting*.

The order of convergence of an iteration method gives an indication of the asymptotic behaviour of the error, i.e., the behaviour after a large number of steps. In practice, one must stop iterating after a finite number of steps. We want to estimate the error in an approximation $\bar{x}$ of a simple root x^*. Using the mean value theorem, we get

$$f(\bar{x}) = f(\bar{x}) - f(x^*) = (\bar{x} - x^*)f'(\xi),$$

where ξ lies between $\bar{x}$ and x^*. Since x^* is a simple root, we have $f'(x^*) \neq 0$, and, if $\bar{x}$ is close to x^*, we also have $f'(\xi) \neq 0$ (f' is assumed to be continuous). Then, we get

$$|\bar{x} - x^*| = \frac{|f(\bar{x})|}{|f'(\xi)|},$$

and, if $|f'(x)| \geq M$ for all x close to x^*, we can make the following estimate

$$|\bar{x} - x^*| \leq \frac{|f(\bar{x})|}{M}.$$

In practice, we compute an approximation $\overline{f}(\overline{x})$ of $f(\overline{x})$. If the absolute error in this approximation is bounded by δ, then $\left|\overline{f}(\overline{x}) - f(\overline{x})\right| \le \delta$, and

$$\left|f(\overline{x})\right| \le \left|\overline{f}(\overline{x})\right| + \delta,$$

which we can use in the estimate. This error estimate is independent of the method we have used to obtain our approximation $\overline{x}$, and, consequently, it is called the method–independent error estimate.

We summarize:

The method–independent error estimate.
Assume that
$$\left|f'(x)\right| \ge M,$$
for all x close to x^*, and that $\overline{x}$ is an approximation of a simple root x^*. Then
$$\left|\overline{x} - x^*\right| \le \frac{\left|\overline{f}(\overline{x})\right| + \delta}{M},$$
where $\overline{f}(\overline{x})$ is an approximation of $f(\overline{x})$, and
$$\left|\overline{f}(\overline{x}) - f(\overline{x})\right| \le \delta.$$

Example 4.4.1   We have solved the equation $f(x) = x - e^{-x} = 0$ with several different methods in the preceding section, and the root appears to be 0.567143 with six correct decimals. We shall now show strictly that this is the case.

Put $\overline{x} = 0.567143$, and compute $\overline{f}(\overline{x}) = -4.551 \cdot 10^{-7}$. If we use a calculator with 10 decimal digits, we can put $\delta = 0.5 \cdot 10^{-10}$ (if the exponential function is computed with ten correct decimals and the subtraction is exact). Further, we have $f'(x) = 1 + e^{-x}$, so that

$$f'(\overline{x}) \approx 1.567.$$

We shall choose M such that $\left|f'(x)\right| \ge M$ close to the root. Here we take $M = 1.5$. Finally, we get

$$\left|\overline{x} - x^*\right| \le \frac{4.55 \cdot 10^{-7} + \frac{1}{2} \cdot 10^{-10}}{1.5} \le 0.4 \cdot 10^{-6},$$

or, equivalently,

$$x^* = 0.567143 \pm 0.4 \cdot 10^{-6}.$$

The accuracy that can be obtained for an approximation of a root x^* depends on the accuracy with which we can compute the function values. We assume that our approximation $\overline{f}(x)$ of the exact function $f(x)$ can be written

$$\overline{f}(x) = f(x) + \delta(x),$$

where $|\delta(x)| \leq \delta$ close to the root. Since δ depends on rounding errors, it is not even continuous in general. In Figure 4.4.2, we have illustrated a function computed in IEEE single precision (not using Horner's rule (see Section 4.5), but a naive algorithm).

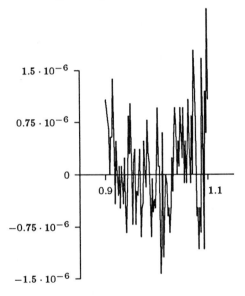

Figure 4.4.2   The polynomial $p(x) = x^6 - 6x^5 + 15x^4 - 20x^3 + 15x^2 - 6x + 1$ computed in IEEE single precision.

From the method–independent error estimate, we see that the best we can achieve is to determine an approximation $\overline{x}$ such that $\overline{f}(\overline{x}) = 0$. Then we have

$$|\overline{x} - x^*| \leq \epsilon = \frac{\delta}{M}.$$

$\epsilon = \dfrac{\delta}{M}$ is called **the attainable accuracy**.

Example 4.4.3   In Example 4.4.1, we have assumed that the calculator computes with 10 decimal digits, and that $\delta = \frac{1}{2} \cdot 10^{-10}$. The attainable accuracy is then

$$\epsilon = \frac{\frac{1}{2} \cdot 10^{-10}}{1.5} = \frac{1}{3} \cdot 10^{-10}.$$

It is obvious that the value of the derivative at the root also affects the attainable accuracy. This is illustrated in Figure 4.4.4.

If the function values are computed with the same accuracy at both roots in Figure 4.4.4, then the root α_2 can be obtained with higher accuracy than α_1, since the derivative has a much larger magnitude at α_2.

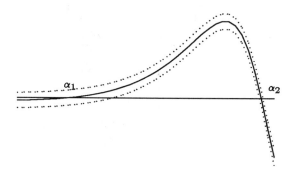

Figure 4.4.4   The root α_1 is ill-conditioned, α_2 is well-conditioned.

The above argument also indicates that multiple roots are ill-conditioned, since the derivative is equal to zero at a multiple root. We determine the attainable accuracy for a double root. Taylor expansion around the double root x^* gives

$$f(x) = f(x^*) + (x - x^*)f'(x^*) + \frac{(x - x^*)^2}{2}f''(\eta)$$
$$= \frac{(x - x^*)^2}{2}f''(\eta).$$

If the maximal absolute error of the function values is δ, we get the attainable accuracy for a double root,

$$\epsilon \approx \sqrt{\frac{2\delta}{|f''(x^*)|}}.$$

The magnitude of the second derivative at x^* is a measure of the curvature at that point. It is seen in Figure 4.4.5 that the attainable accuracy depends on the curvature.

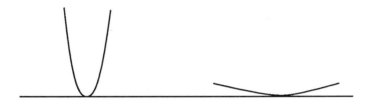

Figure 4.4.5

The following example shows that not only multiple roots can be ill-conditioned.

Example 4.4.6   Consider the polynomial

$$p(x) = (x - 1)(x - 2) \; \cdots \; (x - 20) = x^{20} - 210x^{19} + \cdots + 20!$$

If the coefficient of x^{19} is changed by 2^{-23}, then the roots of the perturbed polynomial become

1.00000	6.00001	$10.09527 \pm 0.64350i$
2.00000	6.99997	$11.79363 \pm 1.65233i$
3.00000	8.00727	$13.99236 \pm 2.51883i$
4.00000	8.91725	$16.73074 \pm 2.81262i$
5.00000	20.84691	$19.50244 \pm 1.94033i$

(the roots are rounded to five decimals).

4.5   Algebraic Equations

An algebraic equation is an equation $p(x) = 0$, where $p(x)$ is a polynomial

$$p(x) = a_0 x^n + a_1 x^{n-1} + \cdots + a_{n-1}x + a_n.$$

The fundamental theorem of algebra states that an nth degree polynomial has exactly n zeros, and if the coefficients $a_0, a_1, \ldots, a_n$ are real, then the zeros ar pairwise conjugate.

Newton–Raphson's method can of course be used also for the solution of algebraic equations. In each iteration, we shall then compute the value of the polynomial and its derivative. It is inefficient to compute each term in the polynomial separately, and then add them up. If the value of x^k is used for the computation of x^{k+1}, this method requires $2n-1$ multiplications and n additions. If, instead, we write the polynomial in the form (for simplicity we here put $n=5$)

$$p(x) = ((((a_0x + a_1)x + a_2)x + a_3)x + a_4)x + a_5,$$

we can compute $p(x_0)$ recursively

$$b_0 = a_0,$$
$$b_k = a_k + b_{k-1}x_0, \qquad k = 1, 2, \ldots, 5,$$
$$p(x_0) = b_5.$$

(In the general case we let $k = 1, 2, \ldots, n$, and have $p(x_0) = b_n$.) Using this method, which is called **Horner's rule**, we can compute $p(x_0)$ with n additions and n multiplications. When the calculations are done by hand, the following scheme is practical:

a_0	a_1	a_2	a_3	a_4	a_5	
	b_0x_0	b_1x_0	b_2x_0	b_3x_0	b_4x_0	x_0
b_0	b_1	b_2	b_3	b_4	b_5	$= p(x_0)$

Example 4.5.1   We let $p(x) = 3x^4 - 2x^2 + x + 1$, and compute $p(2)$.

3	0	-2	1	1	
	6	12	20	42	2
3	6	10	21	43	$= p(2)$

We shall now show that Horner's rule can also be used to compute $p'(x_0)$. The quantities $b_0, b_1, \ldots, b_{n-1}$ in the scheme are the coefficients of the quotient polynomial $q(x)$ that is obtained when we divide $p(x)$ by $x - x_0$:

$$p(x) = (x - x_0)q(x) + r,$$

$$q(x) = b_0x^{n-1} + b_1x^{n-2} + \cdots + b_{n-1}.$$

This can be seen by multiplying together $(x - x_0)q(x)$, and identifying coefficients in the left and right hand sides. We get

$$a_0 = b_0,$$
$$a_1 = b_1 - x_0 b_0,$$
$$a_2 = b_2 - x_0 b_1,$$

$$\vdots$$

$$a_k = b_k - x_0 b_{k-1},$$

$$\vdots$$

$$a_n = r - x_0 b_{n-1},$$

which is exactly the recursion that defines Horner's rule (if we put $r = b_n$).
    Differentiate
$$p'(x) = (x - x_0)q'(x) + q(x),$$
so that
$$p'(x_0) = q(x_0).$$
Thus, we can compute $p'(x_0)$ using the coefficients $b_0, b_1, \ldots, b_{n-1}$ in an analogous scheme.

Example 4.5.2   We compute $p'(x)$ for $p(x) = 3x^4 - 2x^2 + x + 1$, using the scheme from Example 4.5.1.

3	0	−2	1	1		
	6	12	20	42	2	
3	6	10	21	43	$= p(2)$	
	6	24	68	2		
3	12	34	89	$= p'(2)$		

(Check the result by differentiating the polynomial explicitly.)

    The computation of $p(x_0)$ and $p'(x_0)$ can be organized so that no intermediate results need be stored:

$$p := a_0;$$
$$pp := p;$$
for $k := 1$ **to** $n - 1$ **do**
$$\quad p := p * x_0 + a_k;$$
$$\quad pp := pp * x_0 + p;$$
$$p := p * x_0 + a_n;$$

Higher derivatives of $p(x)$ can be computed analogously. Differentiate $p(x) = (x - x_0)q(x) + r$ twice:

$$p''(x) = (x - x_0)q''(x) + 2q'(x),$$

which gives

$$p''(x_0) = 2q'(x_0).$$

To obtain $p''(x_0)$, we therefore need to compute $q'(x_0)$, which we can do by Horner's rule as before.

Example 4.5.3   The second derivative of the polynomial $p(x) = 3x^4 - 2x^2 + x + 1$ is to be computed for $x = 2$. From Example 4.5.2, we have

$$
\begin{array}{ccc|c}
3 & 12 & 34 & \\
 & 6 & 36 & 2 \\
\hline
3 & 18 & 70 & = p''(2)/2
\end{array}
$$

Thus we get $p''(2) = 140$.

It can be seen from the above argument that Horner's rule can be used to divide a polynomial by a linear factor $x - x_0$. This is called **synthetic division**. We have

$$p(x) = (x - x_0)q(x) + p(x_0),$$

and if x_0 is a zero of the polynomial, we have $p(x) = (x - x_0)q(x)$. When a zero x_0 of the polynomial has been determined, e.g., by Newton–Raphson's method, we can divide by $x - x_0$ using Horner's rule, and then continue determining the zeros of the quotient polynomial $q(x)$. This is called **deflation**. A careful rounding error analysis shows that in order to obtain good accuracy for all zeros, one shall remove the ones of small magnitude first.

Example 4.5.4   The polynomial $p(x) = x^3 - 4x^2 + 7x - 4$ has a zero $x = 1$. We factor it out by synthetic division.

$$
\begin{array}{cccc|c}
1 & -4 & 7 & -4 & \\
 & 1 & -3 & 4 & 1 \\
\hline
1 & -3 & 4 & 0 &
\end{array}
$$

Thus,

$$p(x) = (x - 1)(x^2 - 3x + 4).$$

To be sure of convergence for Newton–Raphson's method and secant methods we must have good initial approximations of the roots. If good approximations are not known, then we can use **Laguerre's method**. The iteration formula for this method is

$$z_{k+1} = z_k - \frac{np(z_k)}{p'(z_k) \pm \sqrt{H(z_k)}},$$

where

$$H(z) = (n - 1)[(n - 1)(p'(z))^2 - np(z)p''(z)],$$

and n is the degree of the polynomial. Laguerre's method has cubic convergence for simple roots and good global convergence properties. In contrast to, e.g., Newton–Raphson's method, Laguerre's method can converge to complex roots even if the initial approximation is real.

Example 4.5.5 Let $p(z) = (z^2 + 1)(z^2 + 2)$. With $z_0 = 0$, we get $H(z_0) = -144$, and the following results:

n	z_n
1	$\pm \frac{2}{3} i$
2	$\pm 0.9631540 i$
3	$\pm 0.9998787 i$
4	$\pm 1.000000 i$

4.6 Computer Implementation of $\sqrt{a}$

In this section, we shall describe one possible way of implementing the square root function $\sqrt{a}$ in a computer with binary arithmetic. The aim of this section is to show how to use the theory to derive a practical algorithm for an important application.

The standard for floating point arithmetic, described in Chapter 2, prescribes that the square root function is implemented together with the arithmetic operations. It may happen, however, that for a very simple microprocessor, which is embedded in a larger system, the whole standard is not provided. In such cases, a systems programmer may have to include the standard functions needed.

We assume that the computer has binary arithmetic, and that we want to compute the square root of normalized binary numbers with $t+1$ digits in the significand:

$$A = 1.b_1 b_2 \ldots b_t \cdot 2^e.$$

If the exponent is odd, we shift the significand one step to the left, so that the number is of the form

$$A = a \cdot 2^{2k},$$
$$a = c_1 c_0.d_1 d_2 \ldots d_t.$$

We then have

$$\sqrt{A} = \sqrt{a} \cdot 2^k.$$

The exponent of the result is obtained by shifting one step to the right. Thus, we need a method to compute the square root of a binary number a that satisfies

$$1 \le a < 4.$$

One common method to compute the square root is to use Newton–Raphson's method, which, applied to the equation $f(x) = x^2 - a = 0$, gives the iteration

$$x_{n+1} = \frac{1}{2} \left(x_n + \frac{a}{x_n} \right).$$

We first show a theorem on monotone convergence.

Theorem 4.6.1   For any x_0, $0 < x_0 < \infty$, the square root iteration generates a decreasing sequence

$$x_1 \ge x_2 \ge x_3 \ge \cdots \ge \sqrt{a},$$

that converges to $\sqrt{a}$.

Proof.   The iteration formula gives

$$x_{n+1} - \sqrt{a} = \frac{1}{2} \left(x_n + \frac{a}{x_n} \right) - \sqrt{a}$$

$$= \frac{1}{2x_n} \left(x_n^2 + a - 2\sqrt{a}\, x_n \right) = \frac{1}{2x_n} (x_n - \sqrt{a})^2 \ge 0,$$

which shows that

$$x_n \ge \sqrt{a}, \qquad n = 1, 2, \ldots.$$

Similarly, we get

$$x_n - x_{n+1} = x_n - \frac{1}{2}x_n - \frac{1}{2}\frac{a}{x_n} = \frac{1}{2x_n}(x_n^2 - a) \geq 0.$$

Hence the sequence is decreasing

$$x_1 \geq x_2 \geq \ldots \geq \sqrt{a}.$$

Every monotonically decreasing, bounded sequence is convergent, and therefore we can define $x^* = \lim_{n \to \infty} x_n$. Since

$$x^* = \frac{1}{2}\left(x^* + \frac{a}{x^*}\right),$$

we get (by rewriting that equation) $(x^*)^2 = a$, which means that $x^* = \sqrt{a}$.

Convergence can also be proved using Theorem 4.3.1. We have $\varphi(x) = \frac{1}{2}(x + a/x)$, and

$$0 \leq \varphi'(x) = \frac{1}{2}\left(1 - \frac{a}{x^2}\right) \leq \frac{1}{2},$$

since $x \geq \sqrt{a}$. ∎

The rate of convergence of the square root iteration can be estimated using the formula

$$x_{n+1} - \sqrt{a} = \frac{1}{2x_n}(x_n - \sqrt{a})^2$$

(see the proof above). Since $a \geq 1$, we of course choose $x_0 \geq 1$, and therefore we have $x_n \geq 1$, $n = 0, 1, \ldots$. We can now estimate

$$x_{n+1} - \sqrt{a} \leq \frac{1}{2}\left(x_n - \sqrt{a}\right)^2,$$

which confirms that we have quadratic convergence.

We do not expect to perform many iterations, and therefore we check how the error decreases in the first iterations:

$$x_1 - \sqrt{a} \leq \frac{1}{2}(x_0 - \sqrt{a})^2,$$

$$x_2 - \sqrt{a} \leq \frac{1}{2}(x_1 - \sqrt{a})^2 \leq \frac{1}{2^3}(x_0 - \sqrt{a})^4,$$

$$x_3 - \sqrt{a} \leq \frac{1}{27}(x_0 - \sqrt{a})^8.$$

If we choose a good initial approximation, we will get very rapid convergence. Now, how can one choose good initial approximations that are to be found quickly? Since memory is quite cheap, we can make a table of initial approximations as follows.

The first four bits of a are one of the following 12 combinations:

$$01.00,$$
$$01.01,$$
$$01.10,$$
$$01.11,$$
$$10.00,$$
$$10.01,$$
$$10.10,$$
$$10.11,$$
$$11.00,$$
$$11.01,$$
$$11.10,$$
$$11.11.$$

In a table we store the square roots of

$$c_1 c_0.d_1 d_2 1,$$

where $c_1 c_0.d_1 d_2$ is any of the combinations above. We use the first four bits of a as address in the table. E.g., if $a = (10.01101\dots)_2$, we get from the table the initial approximation $x_0 = \sqrt{(10.011)_2}$. We get the same initial approximation also for $a = (10.0100\dots)_2$. We shall now examine how large the difference $|x_0 - \sqrt{a}|$ can be. The mean value theorem gives

$$|\sqrt{a+h} - \sqrt{a}| = \frac{|h|}{2\sqrt{\xi}} \le \frac{|h|}{2},$$

since we have assumed that $a \ge 1$. The maximal value of h is 2^{-3} (e.g., if $a = (10.000\dots0)_2$, we get $x_0 = \sqrt{(10.001)_2}$). Thus, we have

$$|x_0 - \sqrt{a}| \le 2^{-4}.$$

(In general, if we use the first q bits of a as address in the table we get $|x_0 - \sqrt{a}| \le 2^{-q}$).

We can now directly read off how large the maximal error will be after the first iterations:

$$|x_1 - \sqrt{a}| \le \frac{1}{2} \cdot 2^{-8} < 2 \cdot 10^{-3},$$

$$|x_2 - \sqrt{a}| \le \frac{1}{2} \cdot 2^{-18} < 2 \cdot 10^{-6},$$

$$|x_3 - \sqrt{a}| \le \frac{1}{2} \cdot 2^{-38} < 2 \cdot 10^{-12}.$$

As we have $t + 1$ bits in the significand, we want to compute the square root with an error less than 2^{-t-1}. With $t = 23$ (as in the floating point standard), and the table size we assumed above, we must perform three iterations to get the required accuracy, since $2^{-24} \approx 6 \cdot 10^{-8}$. If, instead, we use six bits as address for the table, we get a table of size 48, and the maximal difference is

$$|x_0 - \sqrt{a}| < 2^{-6} < 0.0157,$$

and

$$|x_2 - \sqrt{a}| < 8 \cdot 10^{-9};$$

here, two iterations are enough to get the required accuracy.

4.7 Systems of Nonlinear Equations

In this section, we briefly treat the problem of solving a system of n nonlinear equations in n unknowns:

$$f_1(x_1, x_2, \ldots, x_n) = 0,$$
$$f_2(x_1, x_2, \ldots, x_n) = 0,$$
$$\vdots$$
$$f_n(x_1, x_2, \ldots, x_n) = 0.$$

We shall use concepts from linear algebra defined in Chapter 8; this section should be read after that chapter.

The system can be written in the usual form

$$f(x) = 0,$$

where $x = (x_1, x_2, \ldots, x_n)^T$ and f is a vector valued function $f(x) = (f_1(x), f_2(x), \ldots, f_n(x))^T$. Several of the methods presented earlier in this chapter can be generalized to systems of nonlinear equations. We first consider Newton–Raphson's method. Let x^0 be an initial approximation of the root x^*. Earlier we derived Newton–Raphson's method by approximating f by the first two terms in the Taylor expansion (see Section 4.2). Analogously, we can here expand each component of f:

$$f_i(x^0 + h) = f_i(x^0) + \sum_{j=1}^{n} \frac{\partial f_i}{\partial x_j}(x^0) \cdot h_j + \ldots, \qquad i = 1, 2, \ldots, n.$$

If we let $J(x)$ denote the $n \times n$ matrix with elements $(\partial f_i / \partial x_j)(x)$, this can be written

$$f(x^0 + h) = f(x^0) + J(x^0)h + \ldots.$$

J is called the **Jacobian** of f. In the same way as before, we determine the correction h by solving the linear system of equations

$$J(x^0)h = -f(x^0),$$

and put

$$x^1 = x^0 + h = x^0 - \left(J(x^0)\right)^{-1} f(x^0).$$

In general, we can write

$$x^{k+1} = x^k - \left(J(x^k)\right)^{-1} f(x^k).$$

The analogy with Newton–Raphson's method in one dimension is obvious. In practice, we do not compute the inverse of J explicitly, but solve the system $Jh = -f(x^k)$ by computing a LU decomposition of J (Gaussian elimination).

Several different variants of Newton–Raphson's method have been devised in order to save work. E.g., it is common to use the same Jacobian for several iterations. At the first iteration, approximately $n^3/3$ operations are needed to compute the correction h, while, if the LU decomposition of J is saved, the following iterations with the same Jacobian require only n^2 operations (cf. Chapter 8). The one-dimensional counterpart of this variant is to approximate the function for several iterations with straight lines that have the same slope, see Figure 4.7.1.

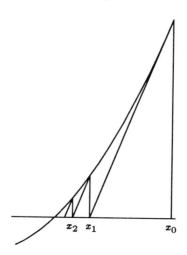

Figure 4.7.1   Newton–Raphson's method with fixed slope.

The analysis of Newton–Raphson's method for systems is more difficult than for a single equation, but one can show that the method converges if x^0 is close enough to x^*, $J(x^*)$ is nonsingular $((J(x^*))^{-1}$ exists), and the second derivatives of f are continuous in a neighborhood of x^*. The convergence is quadratic.

Example 4.7.2 Determine the intersection in the first quadrant between the ellipse

$$f_1(x_1, x_2) = 4x_1^2 + 9x_2^2 - 36 = 0,$$

and the hyperbola

$$f_2(x_1, x_2) = 16x_1^2 - 9x_2^2 - 36 = 0.$$

This system can easily be solved analytically, and we have

$$x_1 = \pm\sqrt{3.6} \approx \pm 1.897367,$$

$$x_2 = \pm\frac{1}{3}\sqrt{21.6} \approx \pm 1.549193.$$

We shall compute the solution using Newton–Raphson's method and the initial approximation $x^0 = (1, 1)^T$. The Jacobian is

$$J(x) = \begin{pmatrix} \dfrac{\partial f_1}{\partial x_1} & \dfrac{\partial f_1}{\partial x_2} \\ \dfrac{\partial f_2}{\partial x_1} & \dfrac{\partial f_2}{\partial x_2} \end{pmatrix} = \begin{pmatrix} 8x_1 & 18x_2 \\ 32x_1 & -18x_2 \end{pmatrix},$$

and the equation $Jh = -f(x)$ can be written

$$\begin{pmatrix} 8x_1 & 18x_2 \\ 32x_1 & -18x_2 \end{pmatrix}\begin{pmatrix} h_1 \\ h_2 \end{pmatrix} = -\begin{pmatrix} f_1(x_1, x_2) \\ f_2(x_1, x_2) \end{pmatrix}.$$

When we put $(x_1, x_2) = (1, 1)$ we get the system

$$\begin{pmatrix} 8 & 18 \\ 32 & -18 \end{pmatrix}\begin{pmatrix} h_1 \\ h_2 \end{pmatrix} = \begin{pmatrix} 23 \\ 29 \end{pmatrix},$$

with the solution

$$\begin{pmatrix} h_1 \\ h_2 \end{pmatrix} = \begin{pmatrix} 1.3 \\ 0.7 \end{pmatrix},$$

and so on. The results are

k	x_1^k	x_2^k
0	1	1
1	2.3	1.7
2	1.93261	1.55588
3	1.89769	1.54921
4	1.89737	1.54919

In the same way as in the one-dimensional case, fixed point iterations can be constructed by rewriting the system in the form

$$x = \varphi(x),$$

where $\varphi(x) = (\varphi_1(x), \varphi_2(x), \ldots, \varphi_n(x))^T$. Starting from an initial approximation, one iterates

$$x^{k+1} = \varphi(x^k), \qquad k = 0, 1, 2, \ldots.$$

Theorem 4.3.1, which gives sufficient conditions for convergence, can be generalized to this case. Assume that $x^* = \varphi(x^*)$, and that the partial derivatives

$$d_{ij}(x) = \frac{\partial \varphi_i}{\partial x_j}, \qquad 1 \leq i, j \leq n,$$

exist for $x \in R = \{x \mid \|x - x^*\| < \rho\}$. Let D be a $n \times n$ matrix with elements $d_{ij}(x)$. A sufficient condition for the fixed point iteration to converge for each $x^0 \in R$ is that, for some matrix norm, we have

$$\|D(x)\| \leq m < 1, \qquad x \in R.$$

If this is satisfied, then we have linear convergence:

$$\left\| x^{k+1} - x^* \right\| \leq m \left\| x^k - x^* \right\|.$$

Example 4.7.3 We shall determine a solution of

$$x_1 = \frac{1}{5} \sin(x_1 + x_2),$$

$$x_2 = \frac{1}{5} \cos(x_1 - x_2),$$

in the neighborhood of $x = (0.1, 0.2)^T$. The matrix D is

$$D = \begin{pmatrix} \frac{1}{5}\cos(x_1 + x_2) & \frac{1}{5}\cos(x_1 + x_2) \\ -\frac{1}{5}\sin(x_1 - x_2) & \frac{1}{5}\sin(x_1 - x_2) \end{pmatrix}.$$

The condition for convergence $\|D\|_\infty \leq m < 1$ is satisfied since

$$\|D\|_\infty \leq \frac{1}{5} + \frac{1}{5} = 0.4.$$

The iteration

$$x_1^{k+1} = \frac{1}{5} \sin(x_1^k + x_2^k),$$

$$x_2^{k+1} = \frac{1}{5} \cos(x_1^k - x_2^k),$$

gives

k	x_1^k	x_2^k
0	0.1	0.2
1	0.059104	0.199001
2	0.051050	0.198046
3	0.049306	0.197843
4	0.048928	0.197798
5	0.048846	0.197788
6	0.048828	0.197787
7	0.048824	0.197785
8	0.048824	0.197785

Exercises

1. Illustrate graphically the solution of the equation $f(x) = x - e^{-x}$ by the secant method for $x_0 = 0.55$ and $x_1 = 0.575$. Solve the equation also with Regula Falsi, and illustrate the solution graphically.

2. Show that Newton–Raphson's method has the asymptotic error constant $\frac{1}{2} f''(x^*)/f'(x^*)$.   *see p. 84*

3. Compute the root of the equation

$$f(x) = \sqrt{x} - e^{-x} = 0,$$

with five correct decimals. Use Newton–Raphson's method.

4. Assume that our computer implements the IEEE single precision standard, and that standard functions are computed with the accuracy

$$\text{fl}[\sqrt{x}] = \sqrt{x}(1 + \epsilon_1),$$
$$\text{fl}[e^{-x}] = e^{-x}(1 + \epsilon_2),$$

where $|\epsilon_i| \le \mu$, $i = 1, 2$ (μ denotes the unit roundoff). How accurately can the equation in Exercise 3 be solved?

5. Let

$$f(x) = \cos(\beta x) - x, \quad \beta = 2.7332 \pm 0.5 \cdot 10^{-4}.$$

How accurately can we solve the equation? Determine the root with an absolute error that does not exceed the attainable accuracy by more than 10%.

6. Determine λ so that the fixed point iteration

$$x_{n+1} = \frac{\lambda x_n + 1 - \sin x_n}{1 + \lambda}$$

converges as fast as possible to the root of the equation

$$1 - x - \sin x = 0.$$

Compute the root with six correct decimals.

7. If a processor does not have hardware division, one can implement division by solving the equation $f(x) = 1/x - a = 0$ using Newton–Raphson's method. On some computers (e.g., Cray X-MP) this method is used for the hardware implementation of division. Analyze this algorithm in the same way as the square root algorithm is analyzed in Section 4.6. In particular, investigate how the number of iterations depends on the table size.

8. Compute the positive root of the equation

$$f(x) = \int_0^x \cos t \, dt - 0.5 = 0$$

with five correct decimals. Hint: Solve the equation $p(x) = 0$, where $p(x)$ is a partial sum of the Maclaurin expansion of $\cos t$ integrated term by term. Then use the method–independent error estimate for the original equation.

9. Show that the fixed point iteration $x_{n+1} = \varphi(x_n)$ does not converge if $|\varphi'(x)| > 1$ in an interval around the required root x^*. This does not necessarily mean that the iteration diverges towards infinity. Try the iteration $x_{n+1} = 1 - \lambda x_n^2$, for $\lambda = 0.7$, 0.9 and 2. Illustrate the iterates as a function of n. For $\lambda = 2$, the iteration exhibits a chaotic behavior.

10. Use Newton–Raphson's method to determine the largest root of the equation $f(x) = 16x^3 - 132x^2 - 12x + 99 = 0$. Compute function values and derivatives with Horner's rule. When the largest root has been computed to six significant digits, perform synthetic division and compute the remaining two roots.

11. The equation $f(x) = x^3 - 7.5x^2 + 18x - 14$ has a double root $x^* = 2$. If we apply Newton–Raphson's method to this equation with $x_0 = 2$, we get quite slow convergence

n	x_n	$f(x_n)$	$f'(x_n)$
0	1	-2.5	6
1	1.42	-0.7	2.77
2	1.67	-0.20	1.30
$\vdots$			
9	1.997	$-1.4 \cdot 10^{-5}$	0.0093

a) Show that the following error estimate holds for double roots

$$|\bar{x} - x^*|^2 \leq \left| \frac{2f(\bar{x})}{Q} \right|,$$

where $|f''(x)| \geq Q$ close to the root.
b) Estimate the error in the approximation x_9.
c) Show that Newton–Raphson's method converges linearly for double roots.

12. The equation $\sin(xy) = y - x$ defines y implicitly as a function of x. The function $y(x)$ has a maximum for $(x, y) \approx (1, 2)$. Show that the coordinates for this maximum can be found by solving the nonlinear system of equations

$$\sin(xy) - y + x = 0,$$
$$y \cos(xy) + 1 = 0.$$

Compute an approximation of the coordinates of the maximum point by iterating once with Newton–Raphson's method.

References

The classical theory for the solution of equations is given in

A. S. Householder, *The Numerical Treatment of a Single Nonlinear Equation*, McGraw–Hill, New York, 1970.

A. Ostrowski, *Solution of Equations and Systems of Equations*, Second edition, Academic Press, New York, 1966.

J. Traub, *Iterative Methods for the Solution of Equations*, Prentice–Hall, Englewood Cliffs, New Jersey, 1964.

A modern, robust variant of the secant method is described in

R. P. Brent, *Algorithms for Minimization without Derivatives*, Prentice–Hall, Englewood Cliffs, New Jersey, 1972.

Wilkinson has studied the perturbation theory for the solution of algebraic equations.

J. H. Wilkinson, *Rounding Errors in Algebraic Processes*, Prentice–Hall, Englewood Cliffs, New Jersey, 1963.

The numerical solution of systems of nonlinear equations is treated in

J. M. Ortega and W. C. Rheinboldt, *Iterative Solution of Nonlinear Equations in Several Variables*, Academic Press, New York, 1970.

5 Interpolation

5.1 Introduction

Assume that function values $f_i = f(x_i)$ are known for $(n + 1)$ different points $x_1, x_2, \ldots, x_{n+1}$. We want to determine a function P such that

$$P(x_i) = f_i, \qquad i = 1, 2, \ldots, n + 1.$$

Such a function P is said to **interpolate** f in the points $x_1, x_2, \ldots, x_{n+1}$.

Many different types of interpolating functions can be used, e.g., polynomials, rational functions, trigonometric functions. We shall restrict ourselves to interpolation by polynomials and functions that are built up from polynomials, so-called spline functions.

An interpolating function can be used to estimate the value of f at a point x. If x lies in the interval formed by $x_1, x_2, \ldots, x_{n+1}$, we speak about **interpolation**, otherwise about **extrapolation**.

We can also use P to approximate the derivative of f at a certain point or to approximate the integral of f over an interval. In these cases, it is useful to approximate f by a polynomial or a spline function, since these functions are easy to differentiate and integrate.

It is not always advisable to determine a function approximating f and require that it have the same value as f at certain points. If only approximations of the function values are known, e.g., from measurements, it is better to use an approximation method that can reduce the influence of measurement errors (see Chapter 9).

5.2   Interpolation by Polynomials

Assume that we know the values of a function f in three points and that we want to construct a polynomial that interpolates f in these points. What is the degree of this polynomial? In general, a straight line cannot interpolate three given points. On the other hand, we can determine a second degree polynomial that passes through the three points, and there are many different third degree polynomials with the same property.

Example 5.2.1   Determine a second degree polynomial that interpolates the values

x_i	f_i
0.5	0.4794
0.6	0.5646
0.7	0.6442

If we write the polynomial in the following form, then the computations will become simple:

$$P(x) = c_0 + c_1(x - 0.5) + c_2(x - 0.5)(x - 0.6).$$

The requirement that $P(x_i) = f_i$ gives a system of equations for determining the coefficients c_0, c_1 and c_2:

$$
\begin{aligned}
c_0 &= 0.4794, \\
c_0 + c_1(0.6 - 0.5) &= 0.5646, \\
c_0 + c_1(0.7 - 0.5) + c_2(0.7 - 0.5)(0.7 - 0.6) &= 0.6442.
\end{aligned}
$$

We compute c_0 from the first equation, use this value in the second equation, and compute c_1, etc. In this way, we get

$$
\begin{aligned}
c_0 &= 0.4794, \\
c_1 &= 0.8520, \\
c_2 &= -0.2800,
\end{aligned}
$$

and

$$P(x) = 0.4794 + 0.852(x - 0.5) - 0.28(x - 0.5)(x - 0.6).$$

Note that the first two terms in P are the equation of a straight line that interpolates f in the points $x = 0.5$ and $x = 0.6$. We then get the second degree polynomial that interpolates f in these points and in $x = 0.7$ by adding a certain second degree polynomial to the linear term. This second degree polynomial is equal to zero for $x = 0.5$ and $x = 0.6$, and the coefficient of the x^2 term of the polynomial is chosen so that $P(0.7)$ attains the required value.

The same technique that was used in the example above to construct a polynomial of higher and higher degree, is used in the proof of the following theorem that is the basis of polynomial interpolation.

Theorem 5.2.2 Let $x_1, x_2, \ldots, x_{n+1}$ be arbitrary, distinct points. For arbitrary values $f_1, f_2, \ldots, f_{n+1}$ there is a *unique* polynomial P of degree $\leq n$ such that

$$P(x_i) = f_i, \qquad i = 1, 2, \ldots, n + 1.$$

Proof. We first prove existence by induction.
For $n = 1$, we can take the polynomial

$$P_1(x) = f_1 + \frac{f_2 - f_1}{x_2 - x_1}(x - x_1),$$

since its degree is at most one, and $P_1(x_1) = f_1, P_1(x_2) = f_2$. Now we make the induction assumption that $P_{k-1}(x)$ is a polynomial of degree $\leq k - 1$ such that

$$P_{k-1}(x_i) = f_i, \qquad i = 1, 2, \ldots, k,$$

and show that there is a polynomial P_k, of degree $\leq k$, that interpolates f in the points $x_1, x_2, \ldots, x_{k+1}$. Put

$$P_k(x) = P_{k-1}(x) + c(x - x_1)(x - x_2) \cdots (x - x_k).$$

The polynomial $P_k(x)$ has degree $\leq k$. Further,

$$P_k(x_i) = f_i, \quad i = 1, 2, \ldots, k.$$

Since the points $x_1, x_2, \ldots, x_{k+1}$ are distinct, we can determine c so that

$$P_k(x_{k+1}) = f_{k+1}.$$

This is satisfied for

$$c = \frac{f_{k+1} - P_{k-1}(x_{k+1})}{(x_{k+1} - x_1)(x_{k+1} - x_2) \cdots (x_{k+1} - x_k)}.$$

Thus, P_k is a polynomial of degree $\leq k$, that interpolates f in x_1, x_2, $\ldots, x_{k+1}$.

In order to show that P is unique, we assume that this is not case: P and Q are different polynomials, both of degree $\leq n$, such that

$$\begin{aligned} P(x_i) &= f_i, \\ Q(x_i) &= f_i, \end{aligned} \qquad i = 1, 2, \ldots, n+1.$$

This means that $P - Q$ is a polynomial of degree less than or equal to n with $n + 1$ different zeros $x_1, x_2, \ldots, x_{n+1}$. Then according to the fundamental theorem of algebra such a polynomial is identically zero, i.e., $P = Q$. ∎

How large will the error be, if the function f is approximated by an interpolating polynomial? To be able to answer this question, we must of course know more about f than its values at some discrete points. For example, if f is not continuous, then the error between the interpolation points can be arbitrarily large.

Theorem 5.2.3 Let f be a function with $n + 1$ continuous derivatives in the interval formed by the points x, x_1, x_2, $\ldots, x_{n+1}$. If P is the polynomial of degree $\leq n$ that satisfies

$$P(x_i) = f(x_i), \qquad i = 1, 2, \cdots, n+1,$$

then

$$f(x) - P(x) = \frac{f^{(n+1)}(\xi(x))}{(n+1)!}(x - x_1)(x - x_2) \cdots (x - x_{n+1}),$$

for some $\xi(x)$ in the interval formed by the points $x, x_1, \ldots, x_{n+1}$.

Proof. For the proof we use Rolle's theorem: If a function g is continuous in the interval $a \leq x \leq b$, and differentiable in $a < x < b$, and $g(a) = g(b)$, then there is at least one point η in (a, b) such that $g'(\eta) = 0$.

We choose an arbitrary point $\bar{x} \neq x_i$, $i = 1, 2, \ldots, n + 1$, and compute the error $f(\bar{x}) - P(\bar{x})$. Assume that, for $x = \bar{x}$, we have

$$f(x) - P(x) = R(x - x_1)(x - x_2) \cdots (x - x_{n+1}).$$

In order to determine the constant R, we introduce the auxiliary function

$$G(x) = f(x) - P(x) - R(x - x_1)(x - x_2) \cdots (x - x_{n+1}).$$

For $x = \bar{x}$, we thus have $G(x) = 0$. Since the polynomial P has been constructed to satisfy $f(x) - P(x) = 0$ for $x = x_1, x_2, \ldots, x_{n+1}$, we have $G(x) = 0$ for these x values also, so that

$$G(x) = 0 \qquad \text{for } x = \bar{x}, x_1, x_2, \ldots, x_{n+1}.$$

According to the assumptions, G is continuously differentiable $(n+1)$ times. Rolle's theorem shows that G' has a zero in each interval between two zeros of G. Therefore, G' has $(n+1)$ zeros. Similarly, in each interval between two zeros of G', there is a zero of G''. Altogether G'' then has n zeros. Repeating this argument, we finally see that $G^{(n+1)}$ has one zero, ξ, in the interval formed by $\bar{x}, x_1, x_2, \ldots, x_{n+1}$:

$$G^{(n+1)}(\xi) = 0.$$

Differentiating the expression for G, we get

$$G^{(n+1)}(x) = f^{(n+1)}(x) - (n + 1)! \cdot R.$$

Thus,

$$R = \frac{f^{(n+1)}(\xi)}{(n + 1)!},$$

where ξ depends on $\bar{x}$. Since $\bar{x}$ is arbitrary, we have proved the theorem. ∎

When a function f is approximated by an interpolating polynomial, the error at the interpolation points is equal to zero. It is tempting to believe that the approximation of f becomes better and better as the number of interpolation points is increased. This is not the case. The classical counterexample was constructed by Runge. He approximated the function

$$f(x) = \frac{1}{1 + 25x^2}$$

on $[-1, 1]$ by polynomials $P_n(x)$ of degree $\leq n$, interpolating f in equidistant points

$$x_i = -1 + (i-1)\frac{2}{n}, \qquad i = 1, \ldots, n+1.$$

In Figure 5.2.4, we illustrate the results for some different values of n.

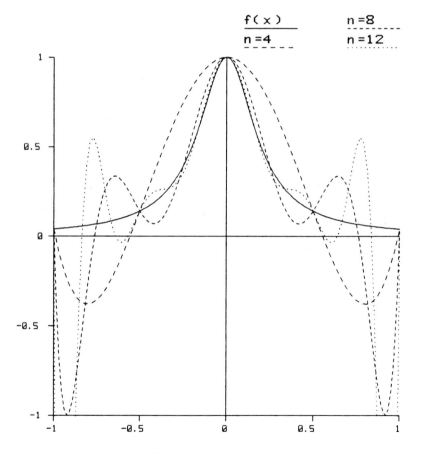

Figure 5.2.4   Runge's example.

The error near the origin is small, but, close to the endpoints -1 and $+1$, the error is increasing with n. For instance $P_{12}(0.96) \approx -3.6$! It can be shown that

$$\lim_{n \to \infty} \left(\max_{-1 \le x \le 1} |f(x) - P_n(x)| \right) = \infty$$

for this function.

In Chapter 9, we shall see that in many cases the error can be made smaller by choosing other interpolation points. However, in general, one cannot be sure that, by choosing suitable interpolation points and a high degree interpolating polynomial, one will obtain a good approximation of a certain function. It can be shown that, for an arbitrary choice of interpolation points, there is a continuous function such that the maximal error in polynomial interpolation tends to infinity as the number of interpolation points tends to infinity.

Conclusion: We should only use polynomial interpolation of low degree.

5.3 Linear Interpolation

In this section, we shall study how different types of errors influence the result in linear interpolation.

The value of the function f at a point $x \in (x_1, x_2)$ is estimated by approximating f by the straight line P through the points (x_1, f_1) and (x_2, f_2):

$$P(x) = f_1 + \frac{x - x_1}{x_2 - x_1}(f_2 - f_1).$$

Example 5.3.1 Find an approximation of $f(0.54)$ when the following function values are known:

x_i	f_i
0.5	0.4794
0.6	0.5646

We get

$$P(0.54) = 0.4794 + \frac{0.54 - 0.5}{0.6 - 0.5}(0.5646 - 0.4794)$$

$$= 0.4794 + \frac{0.04}{0.1} \cdot 0.0852 \doteq 0.5135.$$

(Here we use $\doteq$ to denote correct rounding.)

The approximation of f by a straight line gives rise to a truncation error R_T.

Theorem 5.3.2   If f is twice continuously differentiable, and is approximated for $x_1 \leq x \leq x_2 = x_1 + h$ by linear interpolation between (x_1, f_1) and (x_2, f_2), the truncation error can be estimated as

$$|R_T| \leq \frac{h^2}{8} \max_{x_1 \leq x \leq x_1 + h} |f''(x)|.$$

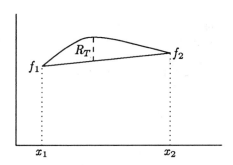

Figure 5.3.3   The truncation error in linear interpolation.

Proof.   According to Theorem 5.2.3,

$$R_T = \frac{f''(\xi(x))}{2!}(x - x_1)(x - x_2),$$

where $\xi(x)$ lies in the interval (x_1, x_2). Put $x = x_1 + ph$, and use the fact that $x_2 = x_1 + h$. Then

$$R_T = \frac{f''(\xi(x_1 + ph))}{2!} h^2 p(p - 1).$$

Since $\max_{0 \leq p \leq 1} |p(p - 1)| = \frac{1}{4}$ the theorem follows.   ∎

In order to use the error bound, we must estimate f'', and we only know the value of f at certain points. The most obvious way to proceed is to approximate f by an interpolating polynomial and differentiate that.

Example 5.3.4 Assume that the function in the preceding example is known at a third point:

$$f(0.7) = 0.6442.$$

In Example 5.2.1, we computed the second degree polynomial that interpolates f in the points 0.5, 0.6 and 0.7:

$$P(x) = 0.4794 + 0.852(x - 0.5) - 0.28(x - 0.5)(x - 0.6).$$

Differentiating, we get

$$P''(x) = -0.28 \cdot 2 = -0.56.$$

Since the distance h between the points is 0.1, we have

$$|R_\mathrm{T}| \lesssim \frac{(0.1)^2}{8} \cdot 0.56 \leq 7 \cdot 10^{-4}.$$

The example has been constructed with $f(x) = \sin x$. In the interval $[0.5, 0.6]$, we thus have $-0.5646 \leq f''(x) \leq -0.4794$, and our method gives a satisfactory estimate of f'' in this case.

Other sources of error that influence the result in linear interpolation are input data errors and rounding errors. Here, the input data are the function argument x, and the function values f_1 and f_2. Let us study how errors in the function values influence the result.

Example 5.3.5 Assume that the function values in Example 5.3.1 have four correct decimals, and estimate how these errors are propagated into the result of the linear interpolation. Let f_i denote the exact function values, and let $\overline{f}_i$ be the given approximations. Put

$$\overline{f}_i - f_i = \epsilon_i, \qquad i = 1, 2.$$

We know that

$$|\epsilon_i| \leq 0.5 \cdot 10^{-4}.$$

The error in the interpolated value can be estimated as

$$R_\mathrm{XF} = [\overline{f}_1 + \frac{x - x_1}{x_2 - x_1}(\overline{f}_2 - \overline{f}_1)] - [f_1 + \frac{x - x_1}{x_2 - x_1}(f_2 - f_1)]$$

$$= \epsilon_1 + \frac{x - x_1}{x_2 - x_1}(\epsilon_2 - \epsilon_1)$$

$$= (1 - \frac{x - x_1}{x_2 - x_1})\epsilon_1 + \frac{x - x_1}{x_2 - x_1}\epsilon_2.$$

We note that the coefficients of ϵ_1 and ϵ_2 are positive since we assumed that $x_1 < x < x_2$. Thus, we get

$$|R_{\mathrm{XF}}| \leq (1 - \frac{x - x_1}{x_2 - x_1}) \cdot 0.5 \cdot 10^{-4} + \frac{x - x_1}{x_2 - x_1} \cdot 0.5 \cdot 10^{-4} = 0.5 \cdot 10^{-4}.$$

To get this estimate of $|R_{\mathrm{XF}}|$, we had to take advantage of the fact that the errors in the two terms of the interpolating polynomial partly cancel. Therefore, we introduced notations for the exact errors ϵ_i in the function values, and simplified R_{XF} as far as possible before we used the known upper bounds for $|\epsilon_i|$.

From the example, we immediately see that the following is valid.

Theorem 5.3.6 If the magnitude of the absolute errors in the function values is at most ϵ, then this results in an error in linear interpolation that can be estimated as

$$|R_{\mathrm{XF}}| \leq \epsilon.$$

In conclusion: the following sources of error affect the result in linear interpolation:

1. An error in the interpolation argument gives rise to R_{XX}.
2. Errors in the function values give rise to R_{XF}.
3. Approximating by a straight line gives R_{T}.
4. Rounding errors in the computations give R_{B}.

R_{XX} and R_{B} can be estimated as was described in Chapter 2.

5.4   Newton's Interpolating Polynomial

Theorem 5.2.2 tells us that the interpolating polynomial of a function is uniquely determined as soon as the points $x_1, x_2, \ldots, x_{n+1}$ are given. In this section, we shall derive an explicit expression for the interpolating polynomial. In fact, there are many different ways to write this polynomial. We start by deriving Newton's interpolating polynomial, which gives simple computations in many practical problems. The derivation is made using the technique of the proof of Theorem 5.2.2.

We start by writing the polynomial in the form

$$P(x) = c_0 + c_1(x-x_1) + c_2(x-x_1)(x-x_2) + \ldots + c_n(x-x_1)(x-x_2)\cdots(x-x_n).$$

The coefficients $c_0, c_1, c_2, \ldots, c_n$ are then determined from the requirement that

$$P(x_i) = f_i, \qquad i = 1, 2, \ldots, n+1.$$

The coefficients of the smallest powers of x can easily be determined from these requirements. For $i = 1, 2$ and 3 we get

$$
\begin{aligned}
c_0 &= f_1, \\
c_0 + c_1(x_2 - x_1) &= f_2, \\
c_0 + c_1(x_3 - x_1) + c_2(x_3 - x_1)(x_3 - x_2) &= f_3.
\end{aligned}
$$

The matrix of this system of equations is lower triangular, and, since $x_i \neq x_k$ for $i \neq k$, all diagonal elements are nonzero. Thus, the coefficients can be computed in order (cf. Section 8.2):

$$c_0 = f_1,$$

$$c_1 = \frac{f_2 - f_1}{x_2 - x_1}, \tag{5.4.1}$$

$$c_2 = \frac{f_3 - f_1 - \frac{x_3 - x_1}{x_2 - x_1}(f_2 - f_1)}{(x_3 - x_1)(x_3 - x_2)}.$$

For the following coefficients the expressions become more complicated. The computations are simplified if we compute the coefficients recursively. To do this, we note how $P(x)$ is structured. The first term is the constant c_0, that interpolates f in x_1. To this term, we add a first degree expression, and we get a straight line that interpolates f in the points x_1 and x_2.

In general $P(x)$ is equal to $P_n(x)$, which is computed as follows:

$$
\begin{aligned}
P_0(x) &= c_0, \\
P_k(x) &= P_{k-1}(x) + c_k(x - x_1)(x - x_2)\cdots(x - x_k), \\
&\qquad k = 1, 2, \ldots, n,
\end{aligned}
\tag{5.4.2}
$$

where P_{k-1} is the polynomial, of degree $\leq k - 1$, that interpolates f in $x_1, x_2, \ldots, x_k$.

From the previously computed expressions for c_0, c_1 and c_2, we see that the values of these coefficients depend on certain function values only. c_0 depends on f_1 only, c_1 depends on f_1 and f_2, c_2 depends on f_1, f_2 and f_3. In general, c_k depends on the function values of f in the points $x_1, x_2, \ldots, x_{k+1}$. Therefore, it is natural to introduce the notation

$$c_k = f[x_1, x_2, \ldots, x_{k+1}].$$

To compute c_k recursively, we observe that, alternatively, $P_k(x)$ can be computed according to

$$P_k(x) = P_{k-1}^{(1)}(x) + \frac{x - x_1}{x_{k+1} - x_1}(P_{k-1}^{(2)}(x) - P_{k-1}^{(1)}(x)), \qquad (5.4.3)$$

where $P_{k-1}^{(1)}$ and $P_{k-1}^{(2)}$ are both polynomials of degree $\leq k - 1$, and

$$P_{k-1}^{(1)}(x) \text{ interpolates } f \text{ in } x_1, x_2, \ldots, x_k,$$
$$P_{k-1}^{(2)}(x) \text{ interpolates } f \text{ in } x_2, x_3, \ldots, x_{k+1}.$$

Let us prove this statement. (5.4.3) defines a polynomial of degree $\leq k$. Does this polynomial interpolate f in $x_1, x_2, \ldots, x_{k+1}$? Putting $x = x_1$, we get:

$$P_k(x_1) = P_{k-1}^{(1)}(x_1) = f_1.$$

For $x = x_i$, $2 \leq i \leq k$, we get

$$P_k(x_i) = P_{k-1}^{(1)}(x_i) + \frac{x_i - x_1}{x_{k+1} - x_1}\left(P_{k-1}^{(2)}(x_i) - P_{k-1}^{(1)}(x_i)\right)$$
$$= P_{k-1}^{(1)}(x_i) = f_i.$$

For $x = x_{k+1}$, we get

$$P_k(x_{k+1}) = P_{k-1}^{(1)}(x_{k+1}) + \frac{x_{k+1} - x_1}{x_{k+1} - x_1}\left(P_{k-1}^{(2)}(x_{k+1}) - P_{k-1}^{(1)}(x_{k+1})\right)$$
$$= P_{k-1}^{(2)}(x_{k+1}) = f_{k+1}.$$

From (5.4.2), we see that $c_k = f[x_1, x_2, \ldots, x_{k+1}]$ is the coefficient of x^k in P_k. Similarly, the coefficient of x^{k-1} in $P_{k-1}^{(1)}$ is $f[x_1, x_2, \ldots, x_k]$, and the coefficient of x^{k-1} in $P_{k-1}^{(2)}$ is $f[x_2, x_3, \ldots, x_{k+1}]$. Using this, and comparing coefficients of x^k in (5.4.3), we get

$$c_k = f[x_1, x_2, \ldots, x_{k+1}]$$
$$= \frac{1}{x_{k+1} - x_1}(f[x_2, x_3, \ldots, x_{k+1}] - f[x_1, x_2, \ldots, x_k]).$$

This formula gives an example of **divided differences**, and it makes it possible to compute c_k recursively.

The computations can be made easily using the following table:

$$
\begin{array}{llll}
x_1 & f_1 \\
& & f[x_1, x_2] \\
x_2 & f_2 & & f[x_1, x_2, x_3] \\
& & f[x_2, x_3] & & f[x_1, x_2, x_3, x_4] \\
x_3 & f_3 & & f[x_2, x_3, x_4] \\
& & f[x_3, x_4] \\
x_4 & f_4
\end{array}
$$

E.g., we have

$$
f[x_1, x_2] = \frac{f_2 - f_1}{x_2 - x_1},
$$

$$
f[x_1, x_2, x_3] = \frac{f[x_2, x_3] - f[x_1, x_2]}{x_3 - x_1},
$$

$$
f[x_1, x_2, x_3, x_4] = \frac{f[x_2, x_3, x_4] - f[x_1, x_2, x_3]}{x_4 - x_1}.
$$

Further, we see that $c_0 = f_1$, $c_1 = f[x_1, x_2]$, $c_2 = f[x_1, x_2, x_3]$, and $c_3 = f[x_1, x_2, x_3, x_4]$.

Obviously, the expressions for c_0 and c_1 are the same as those in (5.4.1). The formula for the computation of c_2 is different, however. If we express c_2 in (5.4.1) using divided differences, we get

$$
c_2 = \frac{1}{x_3 - x_2} \left(\frac{f_3 - f_1}{x_3 - x_1} - \frac{f_2 - f_1}{x_2 - x_1} \right)
$$

$$
= \frac{f[x_1, x_3] - f[x_2, x_1]}{x_3 - x_2} = f[x_2, x_1, x_3],
$$

since $f[x_1, x_2] = f[x_2, x_1]$. We shall show that $f[x_2, x_1, x_3] = f[x_1, x_2, x_3]$. The divided difference $f[x_1, x_2, x_3]$ is the coefficient of the x^2-term in the interpolating polynomial of degree two through the points $(x_1, f_1), (x_2, f_2)$ and (x_3, f_3). Similarly, the divided difference $f[x_2, x_1, x_3]$ is the coefficient of the x^2-term in the interpolating polynomial of degree two through the points $(x_2, f_2), (x_1, f_1)$ and (x_3, f_3). The two polynomials are identical (the order in which the points are taken is of no consequence), and therefore the two divided differences are equal. Using the same argument, we can prove that the value of $f[x_1, x_2, \ldots, x_k]$ does not change if the arguments $x_1, x_2, \ldots, x_k$ are given in a different order.

Example 5.4.4 Determine a third degree polynomial through the points $(-1, 6)$, $(0, 1)$, $(2, 3)$, $(5, 66)$.

We get the following table:

x	$f(x)$	$f[\cdot,\cdot]$	$f[\cdot,\cdot,\cdot]$	$f[\cdot,\cdot,\cdot,\cdot]$
-1	6			
		$\dfrac{1-6}{0-(-1)}=-5$		
0	1		$\dfrac{1-(-5)}{2-(-1)}=2$	
		$\dfrac{3-1}{2-0}=1$		$\dfrac{4-2}{5-(-1)}=\dfrac{1}{3}$
2	3		$\dfrac{21-1}{5-0}=4$	
		$\dfrac{66-3}{5-2}=21$		
5	66			

Thus the polynomial is

$$P(x) = 6 - 5(x+1) + 2(x+1)x + \frac{1}{3}(x+1)x(x-2).$$

Check that this polynomial interpolates the given data!

Definition 5.4.5   **The kth divided difference** of f with respect to the points $x_1, x_2, \ldots, x_{k+1}$ is given recursively by

$$f[x_i] = f(x_i),$$

$$f[x_1, x_2, \ldots, x_{k+1}] = \frac{f[x_2, x_3, \ldots, x_{k+1}] - f[x_1, x_2, \ldots, x_k]}{x_{k+1} - x_1}.$$

Our results are summarized in the following theorem.

Theorem 5.4.6   The polynomial

$$
\begin{aligned}
P_n(x) =\,& f_1 + f[x_1, x_2](x - x_1) + f[x_1, x_2, x_3](x - x_1)(x - x_2) \\
&+ \ldots + f[x_1, x_2, \ldots, x_{n+1}](x - x_1)(x - x_2) \cdots (x - x_n)
\end{aligned}
$$

of degree $\leq n$ satisfies $P_n(x_i) = f_i$, $i = 1, 2, \ldots, n+1$.

This is **Newton's interpolating polynomial.**

In practical applications, the truncation error in polynomial interpolation is often estimated by "the first neglected term". In linear interpolation, this means that the error $f''(\xi(x))(x - x_1)(x - x_2)/2$ is approximated by $f[x_1, x_2, x_3](x - x_1)(x - x_2)$. Is it reasonable to estimate the derivative by a divided difference in this way? Yes, it is, and let us show why. According to (5.4.2) and Theorem 5.4.6, we have

$$P_k(x) = P_{k-1}(x) + f[x_1, x_2, \ldots, x_{k+1}](x - x_1)(x - x_2) \cdots (x - x_k)$$

and

$$P_k(x_{k+1}) = f(x_{k+1}).$$

Therefore,

$$f(x_{k+1}) - P_{k-1}(x_{k+1})$$
$$= f[x_1, x_2, \ldots, x_{k+1}](x_{k+1} - x_1)(x_{k+1} - x_2) \cdots (x_{k+1} - x_k).$$

The right hand side is the truncation error in x_{k+1} when f is approximated by the interpolating polynomial P_{k-1} of degree $\leq k - 1$, with interpolation points $x_1, x_2, \ldots, x_k$. Theorem 5.2.3 gives another expression for this error. If f is k times continuously differentiable, we have

$$f(x_{k+1}) - P_{k-1}(x_{k+1}) = \frac{f^{(k)}(\xi)}{k!}(x_{k+1} - x_1)(x_{k+1} - x_2) \cdots (x_{k+1} - x_k).$$

Comparing the two expressions for the truncation error, we get the following theorem.

Theorem 5.4.7   If the function f is k times continuously differentiable in the interval formed by the points $x_1, x_2, \ldots, x_{k+1}$, then there is a number ξ in this interval, such that

$$f[x_1, x_2, \ldots, x_{k+1}] = \frac{f^{(k)}(\xi)}{k!}.$$

Thus, it is reasonable to estimate the magnitude of $f^{(k)}/k!$ using the divided difference. Note that the divided difference $f[x_1, x_2, \ldots, x_{k+1}]$ is the coefficient of the highest power in $P_k(x)$. In other words, the estimate is based on approximating $f^{(k)}(x)$ by the kth derivative of the interpolating polynomial $P_k(x)$, just as we did in Example 5.3.4 for $k = 2$.

> When a function is approximated by Newton's interpolating polynomial, the truncation error is estimated by the first neglected term.

5.5   Neville's Method

According to (5.4.3) the polynomial P_k, of degree$\leq k$, that interpolates f in $x_1, x_2, \ldots, x_{k+1}$ can be computed using two interpolating polynomials P_{k-1} of degree $\leq k - 1$:

$$P_k(x) = P_{k-1}^{(1)}(x) + \frac{x - x_1}{x_{k+1} - x_1}(P_{k-1}^{(2)}(x) - P_{k-1}^{(1)}(x)).$$

If we compare this with the expression for linear interpolation (Section 5.3) we see that $P_k(x)$ can be thought of as the result of linear interpolation between the two "points" $(x_1, P_{k-1}^{(1)}(x))$ and $(x_{k+1}, P_{k-1}^{(2)}(x))$.

Neville's method is based on this observation: The value of an interpolating polynomial is computed by successive linear interpolations. To describe this method, we introduce some new notation that emphasizes the interpolation points.

Let $P_{12}(x)$ denote the polynomial, of degree ≤ 1, that interpolates the function f in x_1 and x_2. $P_{12}(x)$ can be expressed by a determinant

$$P_{12}(x) = f_1 + \frac{x - x_1}{x_2 - x_1}(f_2 - f_1) = \frac{1}{x_2 - x_1}\begin{vmatrix} x - x_1 & f_1 \\ x - x_2 & f_2 \end{vmatrix}.$$

Let $P_{123}(x)$ be the polynomial, of degree ≤ 2, that interpolates f in the points x_1, x_2 and x_3. This polynomial can be computed by linear interpolation between $(x_1, P_{12}(x))$ and $(x_3, P_{23}(x))$:

$$P_{123}(x) = \frac{1}{x_3 - x_1}\begin{vmatrix} x - x_1 & P_{12}(x) \\ x - x_3 & P_{23}(x) \end{vmatrix}.$$

Let us show this! Since P_{12} and P_{23} are first degree polynomials, the right hand side is a polynomial, of degree ≤ 2. Putting $x = x_1$, we get

$$P_{123}(x_1) = \frac{1}{x_3 - x_1}\begin{vmatrix} 0 & f_1 \\ x_1 - x_3 & P_{23}(x_1) \end{vmatrix} = f_1.$$

Putting $x = x_2$, we get

$$P_{123}(x_2) = \frac{1}{x_3 - x_1}\begin{vmatrix} x_2 - x_1 & f_2 \\ x_2 - x_3 & f_2 \end{vmatrix} = \frac{f_2(x_2 - x_1 - x_2 + x_3)}{x_3 - x_1} = f_2.$$

Finally, putting $x = x_3$, we get

$$P_{123}(x_3) = \frac{1}{x_3 - x_1} \begin{vmatrix} x_3 - x_1 & P_{12}(x_3) \\ 0 & f_3 \end{vmatrix} = f_3.$$

One can continue like this and construct interpolating polynomials of higher degree by successive linear interpolation.

Theorem 5.5.1 Let $P_{i_1 i_2 \ldots i_{k+1}}$ be the polynomial, of degree $\leq k$, that satisfies

$$P_{i_1 i_2 \ldots i_{k+1}}(x_{i_j}) = f_{i_j}, \qquad j = 1, 2, \ldots, k+1.$$

The polynomial can be computed recursively:

$$P_i(x) = f_i,$$

$$P_{i_1 i_2 \ldots i_{k+1}}(x) = \frac{1}{x_{i_{k+1}} - x_{i_1}} \begin{vmatrix} x - x_{i_1} & P_{i_1 i_2 \ldots i_k}(x) \\ x - x_{i_{k+1}} & P_{i_2 i_3 \ldots i_{k+1}}(x) \end{vmatrix}.$$

The theorem is easily proved by induction.

Neville's method is a computational scheme for interpolation based on this theorem:

$$
\begin{array}{c|ccccc}
x - x_1 & x_1 & f_1 = P_1(x) & & & \\
 & & & P_{12}(x) & & \\
x - x_2 & x_2 & f_2 = P_2(x) & & P_{123}(x) & \\
 & & & P_{23}(x) & & P_{1234}(x) \\
x - x_3 & x_3 & f_3 = P_3(x) & & P_{234}(x) & \\
 & & & P_{34}(x) & & \\
x - x_4 & x_4 & f_4 = P_4(x) & & &
\end{array}
$$

Here we have, e.g.,

$$P_{1234}(x) = \frac{1}{x_4 - x_1} \begin{vmatrix} x - x_1 & P_{123}(x) \\ x - x_4 & P_{234}(x) \end{vmatrix}.$$

Example 5.5.2 Compute the value for $x = 5$ of the interpolating polynomial passing through the four points below.

$5-x$	x	$f(x)$

3   2   −8

$$\frac{1}{4-2}\begin{vmatrix} 3 & -8 \\ 1 & 0 \end{vmatrix} = 4$$

1   4   0

$$\frac{1}{6-2}\begin{vmatrix} 3 & 4 \\ -1 & 4 \end{vmatrix} = 4$$

$$\frac{1}{6-4}\begin{vmatrix} 1 & 0 \\ -1 & 8 \end{vmatrix} = 4$$

$$\frac{1}{8-2}\begin{vmatrix} 3 & 4 \\ -3 & -2 \end{vmatrix} = 1$$

−1   6   8

$$\frac{1}{8-4}\begin{vmatrix} 1 & 4 \\ -3 & -20 \end{vmatrix} = -2$$

$$\frac{1}{8-6}\begin{vmatrix} -1 & 8 \\ -3 & 64 \end{vmatrix} = -20$$

−3   8   64

The data were constructed using $f(x) = (x-4)^3$, and the interpolation gives the exact result, as it should.

Neville's method is useful for computing the value of an interpolating polynomial at a particular point. If an explicit expression for the polynomial is needed, then Newton's interpolating polynomial is to be preferred.

5.6   Differences

In the sequel, we consider the case when the interpolation points are equidistant. In order to get simpler computations and, for historic reasons, we introduce the difference operator Δ.

Definition 5.6.1   The **difference operator** Δ maps the sequence $y = \{y_n\}$ onto the sequence $\Delta y = \{y_{n+1} - y_n\}$.
Δy is called **the sequence of first differences of** y.
Sequences of higher order differences are defined recursively:

$$\Delta^k y = \Delta^{k-1}(\Delta y).$$

Example 5.6.2   A sequence y and its sequences of differences:

y	Δy	$\Delta^2 y$	$\Delta^3 y$
1			
	8		
9		8	
	16		0
25		8	
	24		0
49		8	
	32		
81			

In the sequel, we omit the braces {} for the sequence and write

$$\Delta y_n = y_{n+1} - y_n.$$

Further, we shall denote the first element of the sequence Δy by Δy_0, the second element by Δy_1, etc.

Definition 5.6.3 A sequence together with its sequences of differences constitute a **difference scheme**:

$$
\begin{array}{ccccccc}
y_1 \\
& \Delta y_1 \\
y_2 & & \Delta^2 y_1 \\
& \Delta y_2 & & \Delta^3 y_1 \\
y_3 & & \Delta^2 y_2 & & \Delta^4 y_1 \\
& \Delta y_3 & & \Delta^3 y_2 \\
y_4 & & \Delta^2 y_3 \\
& \Delta y_4 \\
y_5
\end{array}
$$

A value in a certain position in the scheme is computed as the difference between the next lower value and the next upper value in the column to the left.

Example 5.6.4

$$\Delta y_1 = y_2 - y_1,$$
$$\Delta y_2 = y_3 - y_2,$$
$$\Delta^2 y_1 = \Delta y_2 - \Delta y_1 = y_3 - 2y_2 + y_1.$$

In general, $\Delta^k y_n$ is a linear combination of $y_n, y_{n+1}, \ldots, y_{n+k}$, i.e., the values that are needed to form the difference.

Now let the sequence be a table of a function f tabulated using the step size h. The sequence $\{1, 9, 25, 49, 81\}$ in Example 5.6.2, e.g., can be considered as a table of the function $f(x) = x^2$ in the points $1+(i-1)h$, $i = 1, 2, \ldots, 5$, with $h = 2$. It is then natural to define the difference operator Δ, acting on a function $f(x)$ tabulated with step size h, as follows:

$$\Delta f(x) = f(x + h) - f(x).$$

The difference operator Δ is a *degrading* operator in the same sense as the differentiation operator: If f is a polynomial of degree n, then Δf is a polynomial of degree $(n - 1)$. Show this as an exercise!

5.7   Equidistant Interpolation Points

In the derivation of Newton's interpolating polynomial, we made no special assumptions concerning the location of the interpolation points; we only assumed that they were distinct. In many applications, the points are equidistant:

$$x_i = x_1 + (i - 1)h, \qquad i = 1, 2, \ldots, n + 1.$$

Then the interpolating polynomial can be written in a simpler form that we shall now derive. Consider the scheme of divided differences in Section 5.4. If the points x_i, $i = 1, 2, 3, 4$, are equidistant, the scheme becomes

$$
\begin{array}{cccccc}
x_1 & f_1 \\
 & & \dfrac{\Delta f_1}{h} \\
x_2 & f_2 & & \dfrac{\Delta^2 f_1}{2h^2} \\
 & & \dfrac{\Delta f_2}{h} & & \dfrac{\Delta^3 f_1}{6h^3} \\
x_3 & f_3 & & \dfrac{\Delta^2 f_2}{2h^2} \\
 & & \dfrac{\Delta f_3}{h} \\
x_4 & f_4
\end{array}
$$

Thus, we get a simple relation between divided differences and Δ differences. Generally, the following can be proved by induction.

$$f[x_1, x_2, \ldots, x_{k+1}] = \frac{\Delta^k f_1}{h^k \cdot k!}. \qquad (5.7.1)$$

Introducing a new variable p by

$$x = x_1 + ph,$$

and using (5.7.1) in Newton's interpolating polynomial, we get

Newton's interpolating polynomial, the equidistant case:

$$f(x_1 + ph) = f_1 + p\,\Delta f_1 + \binom{p}{2}\Delta^2 f_1 + \ldots \binom{p}{m}\Delta^m f_1.$$

The differences $\Delta^i f_1$, $i = 1, 2, \ldots, m$, should be computed directly from a difference scheme without first computing divided differences (see Example 5.7.2).

Also, in this special case of Newton's interpolating polynomial, the truncation error can be estimated by the first neglected term, since, from Theorem 5.4.7,

$$f[x_1, x_2, \ldots, x_{k+1}] = \frac{f^{(k)}(\xi)}{k!}$$

for a point ξ in the interval formed by the points $x_1, x_2, \ldots, x_{k+1}$.

Combining this with (5.7.1), we get

$$\frac{\Delta^k f_1}{h^k} = f^{(k)}(\xi).$$

If $P_{k-1}(x)$ denotes the interpolating polynomial of degree $\leq k - 1$ passing through the equidistant points $x_1, x_2, \ldots, x_k$, we get

$$R_{\mathrm{T}} = f(x) - P_{k-1}(x) = \frac{(x - x_1) \cdots (x - x_k)}{k!} f^{(k)}(\overline{\xi}) =$$

$$= \frac{h^k p(p - 1) \cdots (p - k + 1)}{k!} f^{(k)}(\overline{\xi})$$

$$\approx \frac{p(p - 1) \cdots (p - k + 1)}{k!} \Delta^k f_1.$$

Example 5.7.2   Determine $f(0.54)$ when the following function values are known:

x	0.5	0.6	0.7	0.8
$f(x)$	0.4794	0.5646	0.6442	0.7174

We get the difference scheme (the differences are given in units of 10^{-4}).

x	$f(x)$	Δ	Δ^2	Δ^3
0.5	0.4794			
		852		
0.6	0.5646		−56	
		796		−8
0.7	0.6442		−64	
		732		
0.8	0.7174			

Here, we have $h = 0.1$, $x_1 = 0.5$, so that $p = (x - x_1)/h = 0.4$.
    Newton's interpolating polynomial becomes

$$0.4794 + 0.0852p + \frac{p(p-1)}{2}(-0.0056) + \frac{p(p-1)(p-2)}{6}(-0.0008).$$

For $p = 0.4$, the first three terms give

$$0.4794 + 0.034080 + 0.000672 \doteq 0.5142.$$

The truncation error is estimated from the first neglected term:

$$|R_T| \lesssim 5.2 \cdot 10^{-5}.$$

5.8   Lagrange's Interpolating Polynomial

Lagrange's interpolating polynomial is a different way of representing the interpolating polynomial of a given function. This form is often used for the proof of the Existence Theorem 5.2.2, and also in the derivation of formulas, e.g., for numerical integration.
    As an example, we can write the second degree polynomial $P_2(x)$ through (x_1, f_1), (x_2, f_2) and (x_3, f_3) in the form

$$P_2(x) =$$
$$\frac{(x-x_2)(x-x_3)}{(x_1-x_2)(x_1-x_3)}f_1 + \frac{(x-x_1)(x-x_3)}{(x_2-x_1)(x_2-x_3)}f_2 + \frac{(x-x_1)(x-x_2)}{(x_3-x_1)(x_3-x_2)}f_3.$$

The polynomial in front of f_1 has degree 2 and takes the value 0 for $x = x_2$ and $x = x_3$, and the value 1 for $x = x_1$. The other terms are constructed analogously.

Lagrange's interpolating polynomial, of degree $\leq n$, that interpolates f in the points $x_1, x_2, \ldots, x_{n+1}$ is given by

$$P_n(x) = \sum_{i=1}^{n+1} f_i \cdot p_i(x),$$

where

$$p_i(x) = \frac{(x - x_1) \cdots (x - x_{i-1})(x - x_{i+1}) \cdots (x - x_{n+1})}{(x_i - x_1) \cdots (x_i - x_{i-1})(x_i - x_{i+1}) \cdots (x_i - x_{n+1})}.$$

Show as an exercise that $P_n(x_i) = f_i$ for $i = 1, 2, \ldots, n+1$, and that the degree of the polynomial is $\leq n$.

In practical computation, this formula has the disadvantage that in order to construct an interpolating polynomial of higher degree, the calculations must be done all over again.

5.9 Spline Interpolation

As we have seen, we cannot expect to obtain better approximations by interpolating a given function by polynomials of higher and higher degree. It can also be seen that polynomials are not good for the approximation of functions that change their character in different intervals. This is the case for many functions that describe a physical phenomenon or the shape of an object. Consider, e.g., the following function:

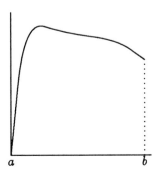

Figure 5.9.1

If we try to approximate such a function by a polynomial, the approximation is bound to be bad for a polynomial of low degree. If we choose a polynomial of high degree, we will get an oscillating approximation. It is better to approximate by different polynomials of low degree in different parts of the interval $[a, b]$.

Assume that we have subdivided the interval $[a, b]$ using points $a \leq x_1 < x_2 < \ldots < x_n \leq b$. The simplest approximation is to use straight lines in each subinterval $[x_i, x_{i+1}]$. The approximating function will be a polygonal curve, i.e., it is not smooth, since its first derivative is discontinuous in $x_2, x_3, \ldots, x_{n-1}$. In spite of this, such an approximation is used in many situations, e.g., in the numerical solution of differential equations by the finite element method, see Chapter 10.

Often third degree polynomials are used in each subinterval. The polynomials are put together in such a manner that the resulting function becomes continuous, and has continuous first and second derivatives.

In general, the function $s(x)$ is called a **spline** of degree $2m+1$, if $s(x)$ is composed of polynomials of degree $2m + 1$ in such a way that the resulting function is $2m$ times continuously differentiable. If $s(x_i) = f(x_i)$, $i = 1, 2, \ldots, n$, then s is an interpolating spline. In the sequel we only treat interpolating splines and therefore we omit "interpolating".

A spline is an elastic wooden ruler, that is fixed in certain points. Such rulers were used, e.g., by ship builders to mark the shape of a ship. The ruler will assume the shape that minimizes its potential energy. Therefore, there can be no large oscillations — the ruler will be as smooth as possible between the points. In the mid forties, some researchers began to describe mathematically the shape of such a ruler. It turned out that the shape can be modelled by an interpolating cubic spline. The potential energy is given by

$$\int_a^b \frac{(f''(x))^2}{(1 + (f'(x))^2)^{3/2}} \, dx.$$

If $(f')^2 \ll 1$ or f' is almost constant, then this expression is minimized by cubic splines.

5.10   Linear Splines

Let $x_1 < x_2 < \ldots < x_n$ be given points, so called **knots**. A **linear spline** is a function $s(x)$ defined on the interval $[x_1, x_n]$ with the following properties

- $s(x)$ is continuous on $[x_1, x_n]$;
- $s(x)$ is a straight line on each subinterval $[x_i, x_{i+1}]$, $i = 1, \ldots, n-1$.

Thus, the function $s(x)$ is composed of $n - 1$ straight lines

$$s(x) = a_i(x - x_i) + b_i, \qquad x_i \leq x \leq x_{i+1}.$$

There are $2(n-1)$ coefficients to be determined. The continuity requirement gives $n-2$ conditions. The remaining n degrees of freedom can, e.g., be used to make the spline take given values in the knots: $s(x_i) = f_i$, $i = 1, 2, \ldots, n$. This gives

$$a_i = \frac{f_{i+1} - f_i}{x_{i+1} - x_i}, \qquad b_i = f_i.$$

If this is satisfied, then the function s is continuous.

An alternative way of representing s is

$$s(x) = \sum_{i=1}^{n} f_i l_i(x),$$

where $l_i(x)$ is the uniquely determined linear spline with the property

$$l_i(x_j) = \delta_{ij} = \begin{cases} 0, & i \neq j, \\ 1, & i = j. \end{cases}$$

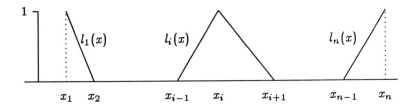

Figure 5.10.1   Basis functions for linear splines.

We have

$$
l_1(x) = \begin{cases} \dfrac{x_2 - x}{x_2 - x_1}, & x_1 \leq x \leq x_2, \\ 0, & x_2 \leq x \leq x_n, \end{cases}
$$

$$
l_i(x) = \begin{cases} 0, & x_1 \leq x \leq x_{i-1}, \\ \dfrac{x - x_{i-1}}{x_i - x_{i-1}}, & x_{i-1} \leq x \leq x_i, \\ \dfrac{x_{i+1} - x}{x_{i+1} - x_i}, & x_i \leq x \leq x_{i+1}, \\ 0, & x_{i+1} \leq x \leq x_n, \end{cases}
$$

$$
l_n(x) = \begin{cases} 0, & x_1 \leq x \leq x_{n-1}, \\ \dfrac{x - x_{n-1}}{x_n - x_{n-1}}, & x_{n-1} \leq x \leq x_n. \end{cases}
$$

The functions $l_i(x)$ are basis functions for all linear splines for a given set of knots. They are so-called **linear B-splines**. Note that these functions are zero outside the open interval (x_{i-1}, x_{i+1}) (see Figure 5.10.1).

Assume that the f_i are chosen equal to $f(x_i)$, where f is a given twice continuously differentiable function. The error in the approximation of f with the linear spline satisfies

$$
f(x) - s(x) = \frac{f''(\xi_i)}{2}(x - x_i)(x - x_{i+1}), \qquad x_i \leq x \leq x_{i+1}.
$$

Let M be an upper bound for the magnitude of $f''(x)$ on the interval $[x_1, x_n]$, and assume that the length of the largest subinterval $[x_i, x_{i+1}]$ is h. Then we can make the estimate

$$
|f(x) - s(x)| \leq \frac{h^2}{8} \cdot M.
$$

The placement of the knots $x_1, x_2, \ldots, x_n$ affects the error, which cannot be seen from the expression above. To make s as good an approximation of f as possible, the knots should be placed closer where f is varying rapidly.

5.11   Cubic Splines

Let $x_1 < x_2 < \ldots < x_n$ be given knots. A **cubic spline** is a function $s(x)$, defined on the interval $[x_1, x_n]$, with the following properties

$s(x), s'(x)$ and $s''(x)$ are continuous on $[x_1, x_n]$;

$s(x)$ is a cubic polynomial on each subinterval $[x_i, x_{i+1}]$.

We shall determine an **interpolating** cubic spline, and require that

$$s(x_i) = f_i, \qquad i = 1, 2, \ldots, n,$$

where f_i are given numbers. Put

$$s(x) = s_i(x), \qquad x_i \le x \le x_{i+1}.$$

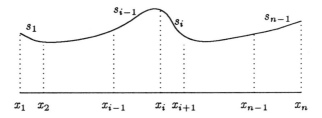

Figure 5.11.1   Cubic spline.

The function s is composed of $(n-1)$ cubic polynomials. Therefore, we have $4(n-1)$ coefficients to determine. Each polynomial shall interpolate f at the endpoint of its interval. This gives $2(n-1)$ conditions. Since the polynomials are put together so that $s_{i-1}(x_i) = s_i(x_i) = f_i$, for $i = 2, \ldots, n-1$, the function $s(x)$ will be continuous at $x_2, \ldots, x_{n-1}$. It remains to satisfy the continuity requirements for s' and s'' in the $(n-2)$ inner knots, i.e.,

$$s_i^{(k)}(x_{i+1}) = s_{i+1}^{(k)}(x_{i+1}), \qquad k = 1, 2, \quad i = 1, 2, \ldots, n-2.$$

This gives $2(n-2)$ conditions. In all, we have $2(n-1) + 2(n-2) = 4n - 6$ conditions to determine $4n - 4$ coefficients.

To get a uniquely determined spline we need two more conditions. E.g., we can require

$$
\begin{aligned}
s''(x_1) &= 0, \\
s''(x_n) &= 0,
\end{aligned}
\qquad \text{``natural'' spline,}
$$

or

$$
\begin{aligned}
s'(x_1) &= f'(x_1), \\
s'(x_n) &= f'(x_n),
\end{aligned}
\qquad \text{``correct'' boundary conditions.}
$$

The term "natural" cubic spline is somewhat misleading in the sense that, in practice, it is often better to use splines with correct boundary conditions, since they give a better approximation of f close to x_1 and x_n (unless $f'' \sim 0$ close to x_1 and x_n). "Natural" alludes to the elastic ruler mentioned in Section 5.9.

We shall now determine an interpolating cubic spline, and write it in the form

$$
s_i(x) = f_i + s_i'(x_i)(x - x_i) + \frac{s_i''(x_i)}{2}(x - x_i)^2 + \frac{s_i^{(3)}(x_i)}{3!}(x - x_i)^3.
$$

(Note that the terms are the same as in a Taylor expansion of s_i around x_i.) Later, we shall also need the first and second derivatives of $s_i(x)$:

$$
s_i'(x) = s_i'(x_i) + s_i''(x_i)(x - x_i) + \frac{s_i^{(3)}(x_i)}{2}(x - x_i)^2,
$$

$$
s_i''(x) = s_i''(x_i) + s_i^{(3)}(x_i)(x - x_i).
$$

We let h_i denote the length of a subinterval

$$
h_i = x_{i+1} - x_i,
$$

and put

$$
\begin{aligned}
z_i &= s_i''(x_i), \qquad i = 1, 2, \ldots, n-1, \\
z_n &= s_{n-1}''(x_n).
\end{aligned}
$$

From the expression for $s_i''(x)$ above, we immediately see that

$$s_i^{(3)}(x_i) = \frac{s_i''(x_{i+1}) - s_i''(x_i)}{h_i}.$$

The requirement that s'' be continuous at the inner knots gives

$$s_i''(x_{i+1}) = s_{i+1}''(x_{i+1}), \qquad i = 1, \dots, n-2.$$

From this,

$$s_i^{(3)}(x_i) = \frac{z_{i+1} - z_i}{h_i},$$

for $i = 1, 2, \dots, n-2$. The relation is valid also for $i = n-1$. Thus, we have

$$s_i(x) = f_i + s_i'(x_i)(x - x_i) + \frac{z_i}{2}(x - x_i)^2 + \frac{z_{i+1} - z_i}{6h_i}(x - x_i)^3,$$

$$i = 1, 2, \dots, n-1. \qquad (5.11.2)$$

The derivative is

$$s_i'(x) = s_i'(x_i) + z_i(x - x_i) + \frac{z_{i+1} - z_i}{2h_i}(x - x_i)^2. \qquad (5.11.3)$$

It remains to determine $s_i'(x_i)$ and z_i. The requirement that $s_i(x_i) = f_i$ is satisfied directly due to our choice of expression for $s_i(x)$. The requirement that $s_i(x_{i+1}) = f_{i+1}$ gives

$$f_i + s_i'(x_i)h_i + \frac{z_i}{2}h_i^2 + \frac{z_{i+1} - z_i}{6h_i}h_i^3 = f_{i+1},$$

or

$$s_i'(x_i) = \frac{f_{i+1} - f_i}{h_i} - z_{i+1}\frac{h_i}{6} - z_i\frac{h_i}{3}, \quad i = 1, 2, \dots, n-1. \quad (5.11.4)$$

To determine the z_i, we use the requirement that $s'(x)$ shall be continuous:
$$s_i'(x_{i+1}) = s_{i+1}'(x_{i+1}), \qquad i = 1, 2, \ldots, n-2.$$

Use (5.11.3) to compute $s_i'(x_{i+1})$. Then, we get

$$s_i'(x_i) + z_i \cdot h_i + \frac{z_{i+1} - z_i}{2h_i} \cdot h_i^2 = s_{i+1}'(x_{i+1}).$$

Replace $s_i'(x_i)$ and $s_{i+1}'(x_{i+1})$ using (5.11.4):

$$\frac{f_{i+1} - f_i}{h_i} - z_{i+1} \cdot \frac{h_i}{6} - z_i \cdot \frac{h_i}{3} + z_i \cdot h_i + \frac{z_{i+1} - z_i}{2} \cdot h_i$$
$$= \frac{f_{i+2} - f_{i+1}}{h_{i+1}} - z_{i+2} \cdot \frac{h_{i+1}}{6} - z_{i+1} \cdot \frac{h_{i+1}}{3}.$$

Simplifying this, we get

$$h_i z_i + 2(h_i + h_{i+1})z_{i+1} + h_{i+1}z_{i+2} = 6\left(\frac{f_{i+2} - f_{i+1}}{h_{i+1}} - \frac{f_{i+1} - f_i}{h_i}\right),$$
$$i = 1, 2, \ldots, n-2.$$

For a natural cubic spline, we have $z_1 = z_n = 0$, and $(n-2)$ equations for determining the $(n-2)$ parameters $z_2, z_3, \ldots, z_{n-1}$. The matrix of the linear system of equations is tridiagonal and symmetric. Since, further, the matrix is diagonally dominant, we can perform Gaussian elimination without pivoting (see Section 8.4).

Correct boundary conditions lead to two extra equations

$$\begin{cases} -\dfrac{h_1}{3}z_1 - \dfrac{h_1}{6}z_2 = \dfrac{f_1 - f_2}{h_1} + f'(x_1), \\ \dfrac{h_{n-1}}{6}z_{n-1} + \dfrac{h_{n-1}}{3}z_n = \dfrac{f_{n-1} - f_n}{h_{n-1}} + f'(x_n). \end{cases}$$

Again, we have a linear system of equations with a tridiagonal, symmetric, and diagonally dominant matrix.

When the z_i have been determined, we can compute $s_i(x)$ from (5.11.2). First (5.11.4) is used for computing $s_i'(x_i)$, and then the value of s_i for a certain x can be computed easily using Horner's scheme (cf. Section 4.5).

Cubic splines can also be represented using basis functions that are zero outside an interval $[x_{i-2}, x_{i+2}]$, so-called cubic B-splines. However, this is beyond the scope of this book.

We summarize:

A cubic spline s that interpolates a given function f at $x_1, x_2, \ldots, x_n$ is given by

$$s(x) = f_i + s_i'(x_i)(x - x_i) + \frac{z_i}{2}(x - x_i)^2 + \frac{z_{i+1} - z_i}{6h_i}(x - x_i)^3,$$

where $x_i \leq x \leq x_{i+1}$, and $h_i = x_{i+1} - x_i$, $i = 1, 2, \ldots, n-1$, and where z_i and $s_i'(x_i)$ are determined as follows:

The quantities z_i, $i = 1, 2, \ldots, n$, satisfy the system of equations

$$h_i z_i + 2(h_i + h_{i+1})z_{i+1} + h_{i+1}z_{i+2}$$
$$= 6\left(\frac{f_{i+2} - f_{i+1}}{h_{i+1}} - \frac{f_{i+1} - f_i}{h_i}\right), \quad i = 1, 2, \ldots, n-2,$$

with additional requirements

$$z_1 = z_n = 0; \qquad \text{(natural spline)}$$

alternatively,

$$\begin{cases} -\dfrac{h_1}{3}z_1 - \dfrac{h_1}{6}z_2 = \dfrac{f_1 - f_2}{h_1} + f'(x_1), & \text{(correct bound-} \\ \dfrac{h_{n-1}}{6}z_{n-1} + \dfrac{h_{n-1}}{3}z_n = \dfrac{f_{n-1} - f_n}{h_{n-1}} + f'(x_n). & \text{ary conditions)} \end{cases}$$

For $s_i'(x_i)$, we have

$$s_i'(x_i) = \frac{f_{i+1} - f_i}{h_i} - z_{i+1}\frac{h_i}{6} - z_i \cdot \frac{h_i}{3}, \quad i = 1, 2, \ldots, n-1.$$

We mentioned earlier that cubic splines are smooth between the knots. This property is expressed in the following theorem.

Theorem 5.11.5 Among all functions g that are twice continuously differentiable on $[a, b]$, and that interpolate f in the points $a = x_1 < x_2 < \ldots < x_n = b$, the interpolating natural cubic spline minimizes

$$\int_a^b (g''(x))^2 \, dx.$$

This means that interpolating cubic splines cannot have large oscillations, since the first derivative of an oscillating function takes large positive and negative values. From the mean value theorem, we then see that the second derivative also takes large values, and therefore the integral cannot be small. Theorem 5.11.5 can be proved using the following lemma.

Lemma 5.11.6 Let s be a natural cubic spline that interpolates a given function f in the knots

$$a = x_1 < x_2 < \ldots < x_n = b.$$

For each function g that is twice continuously differentiable on $[a, b]$, and that interpolates f in the knots, the following is valid:

$$\int_a^b (g''(x))^2 \, dx = \int_a^b (s''(x))^2 \, dx + \int_a^b (g''(x) - s''(x))^2 \, dx.$$

Proof. It is easily seen that

$$\int_a^b \left(g''(x) - s''(x)\right)^2 \, dx = \int_a^b (g''(x))^2 \, dx - \int_a^b (s''(x))^2 \, dx$$
$$- 2 \int_a^b \left(g''(x) - s''(x)\right) s''(x) \, dx.$$

It remains to be proven that the last integral is equal to zero. Consider the interval $[x_i, x_{i+1}]$, and integrate partially

$$I_i = \int_{x_i}^{x_{i+1}} \left(g''(x) - s''(x) \right) s''(x) \, dx$$

$$= \left[\left(g'(x) - s'(x) \right) s''(x) \right]_{x_i}^{x_{i+1}} - \int_{x_i}^{x_{i+1}} \left(g'(x) - s'(x) \right) s^{(3)}(x) \, dx.$$

Since s is a third degree polynomial in the interval, $s^{(3)}$ is constant there. Therefore,

$$\int_{x_i}^{x_{i+1}} \left(g'(x) - s'(x) \right) s^{(3)}(x) \, dx = s^{(3)}(x_i) \int_{x_i}^{x_{i+1}} \left(g'(x) - s'(x) \right) dx$$

$$= s^{(3)}(x_i) \left[\left(g(x) - s(x) \right) \right]_{x_i}^{x_{i+1}}.$$

This expression is equal to zero, since, from the assumption,

$$g(x_i) = s(x_i) = f_i$$

in all the knots.
  We get

$$\int_a^b \left(g''(x) - s''(x) \right) s''(x) \, dx = \sum_{i=1}^{n-1} I_i$$

$$= \sum_{i=1}^{n-1} \left[\left(g'(x_{i+1}) - s'(x_{i+1}) \right) s''(x_{i+1}) - \left(g'(x_i) - s'(x_i) \right) s''(x_i) \right]$$

$$= \left(g'(x_n) - s'(x_n) \right) s''(x_n) - \left(g'(x_1) - s'(x_1) \right) s''(x_1).$$

This expression is equal to zero, since, from the assumption,

$$s''(x_1) = 0 \text{ and } s''(x_n) = 0.$$

∎

Show as an exercise that Theorem 5.11.5 follows from the lemma!
  From the proof, we see that the corresponding theorem is valid also for interpolating cubic splines that satisfy

$$s'(x_1) = f'(x_1) \text{ and } s'(x_n) = f'(x_n),$$

provided that the functions g are such that

$$g'(x_1) = f'(x_1) \text{ and } g'(x_n) = f'(x_n).$$

The following theorem gives information about the approximating properties of cubic splines.

Theorem 5.11.7   Let s be a cubic spline that interpolates a four times continuously differentiable function f at the points

$$a = x_1 < x_2 < \ldots < x_n = b,$$

with correct boundary conditions. Then

$$\max_{a \le x \le b} |s^{(r)}(x) - f^{(r)}(x)| < K_r(\beta)Mh^{4-r}, \qquad r = 0, 1, 2, 3,$$

where $\beta = \max_{i,j} \left(\dfrac{h_i}{h_j} \right)$, $h = \max_i h_i$, $|f^{(4)}(x)| \le M$, $a \le x \le b$, and

$$K_0(\beta) = \tfrac{5}{384}, \qquad K_1(\beta) = \tfrac{1}{216}(9 + \sqrt{3}),$$
$$K_2(\beta) = \tfrac{1}{12}(1 + 3\beta), \qquad K_3(\beta) = \tfrac{1}{2}(1 + \beta^2).$$

The theorem shows that the interpolating spline and its derivatives, up to the third, converge uniformly to f and its derivatives as the largest step size h goes to zero. For equidistant knots, we have $\beta = 1$, and, for non-equidistant knots, β increases when the ratio between the largest and smallest subinterval grows.

For interpolation with splines with correct boundary conditions, we have, in particular, $|s(x) - f(x)| = O(h^4)$; for interpolation with natural splines, on the other hand, we have $|s(x) - f(x)| = O(h^2)$ close to x_1 and x_n.

Exercises

1. We want to compute $f(a) = \sqrt{a}$ for $1 < a < 4$, and we have very high requirements concerning speed.

a) One possible method is to interpolate linearly in an equidistant table. Which table size is needed if we require that $R_{XF} + R_T$ shall be smaller than 2μ? The computer is using the floating point system $(2, 23, -126, 127)$.

b) Another method is to perform one iteration with Newton–Raphson's method applied to the equation $f(x) = x^2 - a = 0$. The starting approximation is taken from a table (see Section 4.6). Which table size is needed if we require that the error after one iteration is smaller than 2μ?

c) The computational work is approximately the same in a) and b). Which method required the smallest table?

2. The following tables with correctly rounded values are given

x	$\sin x$	$\cos x$	$\cot x$
0.001	0.001000	1.000000	1000.0
0.002	0.002000	0.999998	499.999
0.003	0.003000	0.999996	333.332
0.004	0.004000	0.999992	249.999
0.005	0.005000	0.999988	199.998

Compute $\cot(0.0015)$ as accurately as possible:

a) by interpolation in the table for $\cot x$.
b) by interpolation in the tables for $\sin x$ and $\cos x$.
c) Estimate the error in b).
d) Explain the difference between the results in a) and b).

The argument 0.0015 is assumed to be exact.

3. Show that the value of the kth divided difference $f[x_1, x_2, \ldots, x_{k+1}]$ is independent of the order of the points $x_1, x_2, \ldots, x_{k+1}$.

4. Assume that a function is tabulated in the points
$$x_i = x_1 + (i - 1)h, \qquad i = 1, 2, \ldots, 6.$$
The values are correctly rounded to d decimals. Give an upper bound for the maximal error in $\Delta^5 f(x_1)$.

5. Derive a method for estimating $\int_a^b f(x)\,dx$ by interpolating f by a linear spline with the knots $x_i = a + (i - 1)(b - a)/(n - 1)$, $i = 1, 2, \ldots, n$.

6. Show that the interpolating linear spline s with knots $x_1, x_2, \ldots, x_n$ is the function that minimizes
$$\int_{x_1}^{x_n} (g'(x))^2\,dx,$$

among all functions g such that $g(x_i) = f_i$, $i = 1, 2, \ldots, n$, and such that $\int_{x_1}^{x_n} (g'(x))^2 \, dx$ is bounded.

7. Compute an approximation of $f(2.5)$ by interpolating the following data using a natural cubic spline:

x	1	2	3	4
$f(x)$	0	3	4	4

8. a) Determine the natural cubic spline that interpolates the values

x	0	1	2	3
$f(x)$	1	0	0.5	1

   b) Compute $s(0.5)$, $s(1.5)$ and $s(2.5)$, and sketch $s(x)$. Alternatively, use a graphics program to plot $s(x)$.
   c) Compute $s'(3)$.

References

Much of the work in this area was made by Newton, as is also indicated in the names of the interpolation formulas. See, e.g.,

H. H. Goldstine, *A History of Numerical Analysis from the 16th through the 19th Century*, Springer Verlag, 1977.

There is an extensive classical theory, where different representations of interpolating polynomials are derived using operator calculus, see, e.g.,

C.-E. Fröberg, *Numerical Mathematics, Theory and Computer Applications*, The Benjamin/Cummings Publishing Company, Menlo Park, 1985.

A practically oriented presentation of splines is given in

C. deBoor, *A Practical Guide to Splines*, Springer Verlag, New York, 1978.

The theoretical aspects of splines are thoroughly discussed in

L. L. Schumaker, *Spline Functions: Basic Theory*, New York Wiley Corp., New York, 1981.

6   Differentiation and Richardson Extrapolation

6.1 Introduction

We want to compute numerically derivatives of a function that is only known at certain discrete points. To do this we may, e.g., determine an interpolating spline and differentiate that. This method is to be preferred when derivatives are to be computed for many values of the argument, or when the points where the function is known are not equidistant.

In the rest of this chapter, however, we assume that the function values are known at equidistant points, and that the approximate derivative values are needed only at a small number of points. In this case, it is simpler to interpolate the function by a polynomial of low degree, and differentiate this polynomial. We shall see that this is equivalent to approximating derivatives by difference quotients. As a matter of fact, we have already used this technique. In Chapter 5, we estimated the truncation error with the first neglected term, i.e., the derivative $f^{(k)} = d^k f/dx^k$ was estimated by the difference $\Delta^k f/h^k$ (cf. Theorem 5.4.7 and Formula (5.7.1)).

When we have derived different difference approximations to derivatives, we shall use such an approximation to illustrate Richardson extrapolation. With Richardson extrapolation, we can reduce the truncation error in the difference approximation. Richardson extrapolation is a very powerful technique, which we shall later use also in connection with numerical methods for integration and the numerical solution of ordinary differential equations.

Before continuing, we would like to recall the "big O" concept, which will be used in the following chapters. The notation $f(h) = O(h^p)$ as $h \to 0$

means that there are numbers K and h_0 such that $|f(h)| \leq K \cdot h^p$ for all h with $|h| \leq h_0$. We usually write just $f(h) = O(h^p)$, and presuppose "as $h \to 0$."

6.2 Difference Approximations to Derivatives

Assume that the function f is known at the points $x - h, x$ and $x + h$. We want to compute an approximation to $f'(x)$.

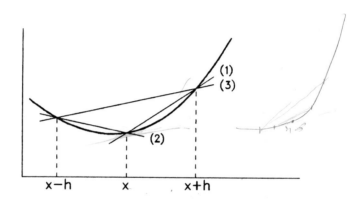

Figure 6.2.1   Approximations to $f'(x)$.

We shall estimate $f'(x)$, i.e., the slope of the tangent of the curve $y = f(x)$ at the point x, using the slope of different straight lines (see Figure 6.2.1). If we use the straight line (1) through $(x, f(x))$ and $(x+h, f(x+h))$, we get

$$f'(x) \approx \frac{f(x + h) - f(x)}{h}.$$

Here, the derivative is approximated by a so-called **forward difference**. We use the notation $D_+(h)$ for this difference quotient.

If, on the other hand, we use the straight line (2) through the points $(x, f(x))$ and $(x - h, f(x - h))$, we get

$$f'(x) \approx \frac{f(x) - f(x - h)}{h}.$$

The derivative is approximated by a so-called **backward difference**, which is denoted $D_-(h)$.

It should be possible to get a better approximation to the derivative by using function values in points that are symmetric around x. Therefore, use the straight line (3) through the points $(x-h, f(x-h))$ and $(x+h, f(x+h))$ in Figure 6.2.1. Then, we get

$$f'(x) \approx \frac{f(x+h) - f(x-h)}{2h},$$

i.e., a **central difference** approximation to the derivative. We denote it $D_0(h)$.

Similarly, we can approximate higher derivatives. For instance, if f is approximated by a second degree interpolating polynomial through the points $(x-h, f(x-h)), (x, f(x))$ and $(x+h, f(x+h))$, and this polynomial is differentiated twice, one gets

$$f''(x) \approx \frac{f(x+h) - 2f(x) + f(x-h)}{h^2}.$$

6.3   The Error in Difference Approximations

When the derivative of a function is approximated by a difference quotient, there is a truncation error R_T. This error can be estimated using a Taylor expansion of the function.

For a twice continuously differentiable function f, whose derivative is approximated by a forward difference, we get

$$\begin{aligned} R_T &= \frac{1}{h}\big(f(x+h) - f(x)\big) - f'(x) \\ &= \frac{1}{h}\Big(f(x) + hf'(x) + \frac{h^2}{2}f''(\xi) - f(x)\Big) - f'(x) \\ &= \frac{h}{2}f''(\xi). \end{aligned}$$

ξ is a point in the interval $(x, x+h)$. We know neither ξ nor f'', but we see that the truncation error is $O(h)$ as $h \to 0$. If f'' does not vary very much in the interval in question, we can therefore reduce R_T to half its size by halving the step size h.

Example 6.3.1   Approximate the derivative of $f(x) = e^x$ for $x = 1$ using the difference quotient

$$D_+(h) = \frac{f(1+h) - f(1)}{h}.$$

For some different values of h, we get:

h	$D_+(h)$	$D_+(h) - e$
0.4	3.3423	$6.24 \cdot 10^{-1}$
0.2	3.0092	$2.91 \cdot 10^{-1}$
0.1	2.8588	$1.41 \cdot 10^{-1}$
0.05	2.7874	$6.91 \cdot 10^{-2}$

We see that the error $D_+(h) - e$ is halved when h is reduced by a factor of two.

If we keep some more terms in the Taylor expansion of the truncation error, we get

$$R_T = \frac{f''(x)}{2} h + \frac{f^{(3)}(x)}{3!} h^2 + \frac{f^{(4)}(x)}{4!} h^3 + \ldots$$

$$= a_1 h + a_2 h^2 + a_3 h^3 + \ldots .$$

We do not know the values of the coefficients $a_k = f^{(k+1)}(x)/(k+1)!$. Later, in connection with Richardson extrapolation, we shall see how we can take advantage of our knowledge of the existence of such an expansion of R_T.

Analogously, we can use Taylor expansions to derive expressions for the truncation error for other difference approximations to derivatives. For instance, we get

$$\frac{f(x+h) - f(x-h)}{2h} = f'(x) + b_1 h^2 + b_2 h^4 + b_3 h^6 + \ldots ,$$

$$\frac{f(x+h) - 2f(x) + f(x-h)}{h^2} = f''(x) + c_1 h^2 + c_2 h^4 + c_3 h^6 + \ldots .$$

Here b_i, c_i, $i = 1, 2, 3, \ldots$, are unknown coefficients, which depend on higher derivatives of f, and which are independent of h.

In particular, we see that these central difference approximations to $f'(x)$ and $f''(x)$ have truncation errors that are $O(h^2)$. Therefore, we can expect them to be more accurate than, e.g., forward difference approximations, provided that the coefficients of the h^2-terms are not large.

Example 6.3.2 Use the central difference

$$D_0(h) = \frac{f(1+h) - f(1-h)}{2h}$$

to approximate the derivative in Example 6.3.1.
We get

h	$D_0(h)$	$D_0(h) - e$
0.4	2.791352	$7.31 \cdot 10^{-2}$
0.2	2.736440	$1.82 \cdot 10^{-2}$
0.1	2.722815	$4.53 \cdot 10^{-3}$
0.05	2.719414	$1.13 \cdot 10^{-3}$

We see that the error $D_0(h) - e$ is reduced approximately by a factor of four when h is halved. Further, the accuracy is considerably better than in Example 6.3.1, where the derivative was approximated using a forward difference.

Since $\lim_{h \to 0} R_T = 0$ for all difference approximations above, it is tempting to believe that it is possible to get an arbitrarily good approximation to a derivative by taking h small enough.

Example 6.3.3 Study the error when the derivative of $f(x) = e^x$ for $x = 1$ is approximated by

$$D_0(h) = \frac{f(1 + h) - f(1 - h)}{2h}.$$

The computations are made on a computer with unit roundoff $\mu = 2^{-27} \approx 7.5 \cdot 10^{-9}$. We start with a step length h that is a power of 10 times the unit roundoff, and then successively we reduce h by a factor of 10. The results are

h	e^{1+h}	e^{1-h}	$D_0(h)$	$D_0(h) - e$
$7.5 \cdot 10^{-1}$	5.726234	1.290387	2.976847	2.59×10^{-1}
$7.5 \cdot 10^{-2}$	2.928545	2.523115	2.720797	2.52×10^{-3}
$7.5 \cdot 10^{-3}$	2.738610	2.698104	2.718306	2.42×10^{-5}
$7.5 \cdot 10^{-4}$	2.720308	2.716257	2.718300	1.82×10^{-5}
$7.5 \cdot 10^{-5}$	2.718484	2.718079	2.718200	-8.18×10^{-5}
$7.5 \cdot 10^{-6}$	2.718302	2.718262	2.718000	-2.82×10^{-4}
$7.5 \cdot 10^{-7}$	2.718284	2.718280	2.720000	1.72×10^{-3}
$7.5 \cdot 10^{-8}$	2.718282	2.718282	2.800000	8.17×10^{-2}
$7.5 \cdot 10^{-9}$	2.718282	2.718282	4.000000	1.28

When h is decreased from $7.5 \cdot 10^{-1}$ to $7.5 \cdot 10^{-3}$, the error decreases by a factor of 100 in each step, as expected. When we go from $h = 7.5 \cdot 10^{-3}$ to $h = 7.5 \cdot 10^{-4}$, the error is decreased only marginally. Then the magnitude of the error increases when h is made smaller. For $h = 7.5 \cdot 10^{-9}$, the result has a relative error of about 50%. This is due to the fact that e^{1+h} and e^{1-h} are not computed accurately enough. The two function values differ by less than 10^{-7} for this value of h. We get a large error due to cancellation.

Thus, when a derivative is approximated by a difference quotient, the result is affected not only by the truncation error, but also by the errors in the function values that are used.

Let us illustrate this using the central difference formula

$$f'(x) \approx \frac{f(x+h) - f(x-h)}{2h}.$$

We only know approximations $\bar{f}(x \pm h)$ to the function values. If

$$|\bar{f}(x \pm h) - f(x \pm h)| \leq \epsilon,$$

the resulting error in the computed difference approximation $\overline{D}_0(h)$ can be estimated as

$$|R_{\mathrm{XF}}| = |\overline{D}_0(h) - D_0(h)| \leq \frac{2\epsilon}{2h} = \frac{\epsilon}{h}.$$

Note that this error *increases* with decreasing step size h. In order to minimize the total error, we shall therefore choose h so that the sum of the truncation error R_{T} and the error R_{XF}, due to errors in the function values, is minimized. Figure 6.3.4 illustrates the typical behaviour of the two different types of errors.

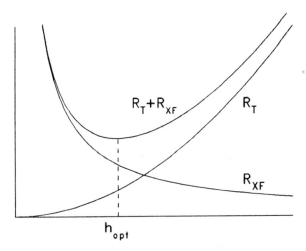

Figure 6.3.4   The errors in difference approximations to derivatives.

The truncation error of the central difference formula is $h^2 f^{(3)}(\xi)/6$ (Exercise 1 at the end of this chapter). To get a small total error we shall therefore minimize

$$\frac{h^2}{6}|f^{(3)}(\xi)| + \frac{\epsilon}{h}$$

by choosing a suitable value of h. Since $f^{(3)}$ is unknown, this is not trivial. (Methods for choosing a suitable value of h have been developed, however. In principle, they are based on estimating $f^{(3)}$ also by a difference quotient.)

In order to determine an optimal h, we shall, in the next section, compute difference approximations for different values of h, and use these to get information about the accuracy of the difference approximations, and also to obtain better accuracy.

6.4 Richardson Extrapolation

Assume that a function $F(h)$ can be computed for several values of $h \neq 0$, and that we want to find the limit of the function as $h \to 0$. This is the type of problem we have when approximating a derivative of a function f at a point x using a difference quotient. The central difference

$$F(h) = \frac{f(x+h) - f(x-h)}{2h}$$

can be computed for different values of h, and we want to compute

$$\lim_{h \to 0} F(h) = f'(x).$$

If the behaviour of the truncation error as $h \to 0$ is known, it is possible to obtain good approximations of $F(0)$, using the value of $F(h)$ for different values of h, and extrapolation.

We shall use the central difference formula to illustrate the method. In Section 6.3, we saw that

$$F(h) = \frac{f(x+h) - f(x-h)}{2h} = f'(x) + b_1 h^2 + O(h^4).$$

Here, $b_1 = f^{(3)}(x)/6$. Thus, we do not know the value of the coefficient, but we know that it is independent of h.

If F is computed for $2h$, we have

$$F(2h) = f'(x) + 4b_1 h^2 + O(h^4).$$

By subtracting the first equation from the second, we can estimate the term $b_1 h^2$ in the truncation error:

$$F(2h) - F(h) = 3b_1 h^2 + O(h^4),$$

or, equivalently

$$\frac{F(2h) - F(h)}{3} = b_1 h^2 + O(h^4).$$

Using this in the expression for $F(h)$, we get

$$F(h) = f'(x) + \frac{F(2h) - F(h)}{3} + O(h^4),$$

or

$$f'(x) = F(h) + \frac{F(h) - F(2h)}{3} + O(h^4).$$

Thus we have an approximation to $f'(x)$ with a truncation error proportional to h^4 instead of h^2.

Example 6.4.1 When approximating $f'(1)$ for $f(x) = e^x$, using central differences in Example 6.3.2, we got

h	$F(h)$
0.4	2.791352
0.2	2.736440

The truncation error in $F(0.2)$ can be estimated as

$$\Delta/3 = (F(0.2) - F(0.4))/3 = -0.018304,$$

and we get

$$f'(1) \approx F(0.2) + \Delta/3 = 2.718136.$$

The error in this approximation is $-1.46 \cdot 10^{-4}$, while the error in $F(0.2)$ is $1.82 \cdot 10^{-2}$. (Here we have used Δ to denote the difference between two adjacent function values in the table.)

For the generalization of this procedure, we introduce the notation

$$F_1(h) = F(h),$$
$$F_2(h) = F_1(h) + \frac{F_1(h) - F_1(2h)}{3}.$$

Example 6.4.2 We start from the approximations $D_0(h) = F_1(h)$ in Example 6.3.2, and compute new approximations $F_2(h)$ to the derivative for $h = 0.2, 0.1, 0.05$. The following results are obtained.

2h use central diff.

h	$F_1(h)$	$\Delta/3$	$F_2(h)$
0.4	2.791352		
		−0.018304	
0.2	2.736440		2.718136
		−0.004542	
0.1	2.722815		2.718273
		−0.001134	
0.05	2.719414		2.718280

trunc error

Study the table! The truncation error in $F_1(h)$, which is proportional to h^2, ought to be reduced by a factor of four when h is halved. The same thing should also happen to the estimates of the truncation error, $\Delta/3$. We see that theory and practice agree very well.

By forming F_2, we eliminate the h^2-term in the truncation error. Thus, we have

$$F_2(h) = f'(x) + c_4 h^4 + O(h^6).$$

If we have computed $F_2(h)$ for some values of h, then we can similarly estimate $c_4 h^4$:

$$F_2(2h) = f'(x) + c_4(2h)^4 + O(h^6),$$

i.e.,

$$\frac{F_2(2h) - F_2(h)}{15} = c_4 h^4 + O(h^6).$$

$$f(x) = f(x_0) + f'(x_0)(x-x_0) + \frac{f''(x_0)}{2!}(x-x_0)^2 + \cdots$$

Using this estimate of the truncation error in the expression for $F_2(h)$, we get

$$F_2(h) = f'(x) + \frac{F_2(2h) - F_2(h)}{15} + O(h^6),$$

or

$$f'(x) = F_2(h) + \frac{F_2(h) - F_2(2h)}{15} + O(h^6).$$

This means that

$$F_3(h) = F_2(h) + \frac{F_2(h) - F_2(2h)}{15}$$

is an approximation to $f'(x)$ with truncation error $O(h^6)$.

Example 6.4.3 Let us compute $F_3(0.1)$ in the previous example:

h	$F_2(h)$	$\Delta/15$	$F_3(h)$
0.2	2.718136		
		0.000009	
0.1	2.718273		2.718282

$F_3(0.1)$ is an approximation to $F(0)$ with six correct decimals. The best approximation in the F_2 column has the error $F_2(0.1) - e \approx 4.5 \cdot 10^{-3}$. Thus, the extrapolation gives a much better accuracy at a small cost.

The above principle for approximating $F(0)$ can be used more generally when $F(h)$ has been computed for two step lengths h and qh, and one knows that the truncation error in $F(h)$ is proportional to h^p. We get:

If

$$F(h) = F(0) + ch^p + O(h^r), \qquad r > p,$$

with a known p and an unknown c, which is independent of h, then

$$F(h) + \frac{1}{q^p - 1}(F(h) - F(qh)) = F(0) + O(h^r).$$

If we know a complete expansion of the truncation error, we can perform repeated Richardson extrapolation. Assume that

$$F(h) = F(0) + a_1 h^{p_1} + a_2 h^{p_2} + a_3 h^{p_3} + \ldots$$

with known exponents $p_1, p_2, p_3, \ldots$ but unknown $a_1, a_2, a_3, \ldots$. Further, assume that $F(h)$ has been computed for step lengths $q^2 h, qh, h, h/q, \ldots$.
Put $F_1(h) = F(h)$, and form

$$F_{i+1}(h) = F_i(h) + \frac{1}{q^{p_i} - 1}(F_i(h) - F_i(qh)), \qquad i = 1, 2, \ldots.$$

In this extrapolation, the h^{p_i}-term is eliminated from the expansion, i.e.,

$$F_{i+1}(h) = F(0) + b_{i+1} h^{p_{i+1}} + b_{i+2} h^{p_{i+2}} + \ldots.$$

We get the following extrapolation scheme:

F_1	$\Delta/(q^{p_1}-1)$	F_2	$\Delta/(q^{p_2}-1)$	F_3
$F_1(q^2h)$				
$F_1(qh)$		$F_2(qh)$		
$F_1(h)$		$F_2(h)$		$F_3(h)$

The entries are computed row by row, and extrapolations are performed until two values in the same *column* agree to the required accuracy. If h is small enough, then the difference between two adjacent values in the same column gives an upper bound for the truncation error.

The influence of errors in the function values on the result, should, in general, be estimated from a scheme of perturbations. For the special case $q = 2, p_i = 2i$, however, it can be shown that absolute errors in $F(h)$ that are bounded by ϵ cause errors in the entries of the scheme that do not exceed 2ϵ in magnitude.

Example 6.4.4 When we approximate $f'(1)$ for $f(x) = e^x$, using the central difference formula, we get the following extrapolation scheme.

h	$F_1(h)$	$\Delta/3$	$F_2(h)$	$\Delta/15$	$F_3(h)$
0.4	2.791352				
		−0.018304			
0.2	2.736440		2.718136		
		−0.004542		0.000009	
0.1	2.722815		2.718273		2.718282
		−0.001134		0.000000	
0.05	2.719414		2.718280		2.718280

We accept $F_3(0.05) = 2.718280$ as an approximation to $f'(1)$. The truncation error is estimated by

$$|R_T| \le |F_3(0.05) - F_3(0.1)| = 2 \cdot 10^{-6}.$$

The $F_1(h)$-entries have six correct decimals. Therefore the errors in the function values give rise to errors no larger than $2 \cdot 0.5 \cdot 10^{-6}$ in any value in the extrapolation scheme.

We disregard here rounding errors in the computation of the corrections $\Delta/3$ and $\Delta/15$ (in general, they can be made small by using arithmetic that is accurate enough).

Richardson extrapolation can also be considered as extrapolation using an interpolating polynomial for $F(h)$. We conclude the section by demonstrating this in a special case.

Assume that

$$F(h) = a_0 + a_1 h^2 + a_2 h^4 + a_3 h^6 + \ldots,$$

and that the values $F(4h_0), F(2h_0)$ and $F(h_0)$ have been computed. It is natural to approximate F by the polynomial of the form $c_0 + c_1 h^2 + c_2 h^4$ which interpolates F in the three known points. The value of this polynomial for $h = 0$ is used to approximate $F(0)$. It is convenient to use Neville's method here (see Section 5.5). If we put $h^2 = x$, then we shall compute the value, for $x = 0$, of the polynomial $p(x) = c_0 + c_1 x + c_2 x^2$ through the points $((4h_0)^2, F(4h_0)), ((2h_0)^2, F(2h_0))$ and $(h_0^2, F(h_0))$. We get the scheme

$0 - x$	x	$p(x)$			
$-(4h_0)^2$	$(4h_0)^2$	F_{00}			
			F_{11}		
$-(2h_0)^2$	$(2h_0)^2$	F_{10}		F_{22}	
			F_{21}		
$-h_0^2$	h_0^2	F_{20}			

Here $F_{i0} = F(2^{2-i}h_0)$, $i = 0, 1, 2$ are known. The other values are computed according to the rules of Neville's method:

$$F_{11} = \frac{1}{(2h_0)^2 - (4h_0)^2} \begin{vmatrix} -(4h_0)^2 & F_{00} \\ -(2h_0)^2 & F_{10} \end{vmatrix} = F_{10} + \frac{1}{3}(F_{10} - F_{00}),$$

$$F_{21} = \frac{1}{h_0^2 - (2h_0)^2} \begin{vmatrix} -(2h_0)^2 & F_{10} \\ -h_0^2 & F_{20} \end{vmatrix} = F_{20} + \frac{1}{3}(F_{20} - F_{10}),$$

$$F_{22} = \frac{1}{h_0^2 - (4h_0)^2} \begin{vmatrix} -(4h_0)^2 & F_{11} \\ -h_0^2 & F_{21} \end{vmatrix} = F_{21} + \frac{1}{15}(F_{21} - F_{11}).$$

We see that we obtain exactly the same values as we did earlier in Richardson extrapolation.

There are also extrapolation methods based on interpolation with rational functions. In some applications, they give better approximations to $F(0)$ than methods based on interpolation by polynomials.

Exercises

1. a) Show that the truncation error in the approximation of $f'(x)$ by the central difference formula is $h^2 f^{(3)}(\xi)/6$, where $\xi \in (x-h, x+h)$.
   b) Show that the total error in the approximation in a) is minimized for $h = \sqrt[3]{3\epsilon/|f^{(3)}(\xi)|}$, where $\xi \in (x-h, x+h)$, and ϵ is an upper bound for the absolute error in the function values. (Here we neglect that ξ depends on h.)
   c) Compute an optimal step length h for approximating $f'(1)$, when $f(x) = e^x$, and the function values can be computed with a relative error smaller than or equal to the unit roundoff 2^{-27}. Estimate the total error in the approximation of the derivative. Compare the result to Example 6.3.3.

2. Show that the truncation error in the approximation

$$\frac{1}{2h^3}(f(x+2h) - 2f(x+h) + 2f(x-h) - f(x-2h))$$

to $f^{(3)}(x)$ is $R_T = a_1 h^2 + a_2 h^4 + a_3 h^6 + \dots$.

3. The following table of a function $f(x)$ is given:

x	$f(x)$
0.6	0.564642
0.8	0.717356
0.9	0.783327
1	0.841471
1.1	0.891207
1.2	0.932039
1.4	0.985450

Compute approximations to $f'(1)$ and $f''(1)$ using repeated Richardson extrapolation. (We have chosen $f(x) = \sin x$ in this example. How well do your results match the exact values?)

4. The expression

$$\frac{1}{12h}(f(x-2h) - 8f(x-h) + 8f(x+h) - f(x+2h))$$

approximates $f'(x)$ with truncation error $R_T = a_4 h^4 + a_6 h^6 + a_8 h^8 + \dots$. Assume that we know the following correctly rounded function values, and want to compute $f'(0.5)$ as accurately as possible.

x	$f(x)$
0.1	0.000167
0.3	0.004480
0.4	0.010582
0.425	0.012679
0.450	0.015034
0.475	0.017662
0.5	0.020574
0.525	0.023787
0.550	0.027313
0.575	0.031165
0.6	0.035358
0.7	0.055782
0.9	0.116673

Use the above difference approximation and Richardson extrapolation to approximate $f'(0.5)$. Estimate the error in the approximations. Why does it not pay off to use Richardson extrapolation in this case?

5. A car starts from standstill, and its speedometer is read every second during acceleration, giving the following readings.

$t(s)$	0	1	2	3	4	5	6	7	8	9
$v(m/s)$	0	3.61	7.22	10.10	12.50	14.62	16.60	18.06	19.54	20.28

What is the acceleration after $4s$? We assume that the speedometer has an error of less than 1%. Use the table and a central difference to compute an approximation with as little total error as possible.

7 Numerical Integration

7.1 Introduction

Numerical integration or quadrature is the computation of approximations to $\int_a^b f(x)\,dx$. We shall derive methods for this.

When are such methods needed? If the integrand f is known only in discrete points, of course. But even if an explicit formula for f is given, it may be impossible to compute the integral exactly because a primitive function cannot be found. This is the case for $\int_0^1 e^{-x^2}\,dx$ and $\int_0^{\pi/2} \sqrt{1 + \cos^2 x}\ \,dx$.

Finally—even if a primitive function is known, the computation of this function may be so costly, that approximate, numerical methods for computing the integral are to be preferred.

We construct methods for numerical integration by approximating f by a function that can easily be integrated, a polynomial. There are two possible ways to obtain good accuracy:

a) approximating f by a single interpolation polynomial of high degree;
b) approximating f by different polynomials of low degree in small subintervals of the interval of integration.

We shall see that method b) is to be preferred.

7.2 Newton–Cote's Quadrature Formulas

Newton–Cote's quadrature formulas are derived by replacing the integrand

f by an interpolating polynomial and integrating that. The interval (a, b) is divided into n subintervals of length $h = (b - a)/n$ using the points $x_i = a + ih$, $i = 0, 1, \ldots, n$. The integrand f is approximated by the interpolating polynomial P_n of degree n through the points $(x_i, f(x_i))$, $i = 0, 1, \ldots, n$. We get

$$\int_a^b f(x)\, dx = \int_a^b P_n(x)\, dx + R_\mathrm{T}.$$

From Theorem 5.2.3, the truncation error is

$$R_\mathrm{T} = \int_a^b (x - x_0)(x - x_1) \ldots (x - x_n) \frac{f^{(n+1)}(\xi(x))}{(n+1)!}\, dx.$$

So, if the required integral is approximated by the integral of an interpolation polynomial of degree n, we get a formula that is exact for all polynomials of degree less than or equal to n. Further, it is easy to show that the quadrature formula expresses the integral as a linear combination of the function values $f(x_i)$, $i = 0, 1, \ldots, n$ To see this, we represent $P_n(x)$ for a moment by Lagrange's interpolation polynomial

$$P_n(x) = \sum_{i=0}^n p_i(x) \cdot f(x_i),$$

where the $p_i(x)$ are polynomials of degree n (see Section 5.8). We get

$$\int_a^b P_n(x)\, dx = \sum_{i=0}^n \left(\int_a^b p_i(x)\, dx \right) f(x_i) = \sum_{i=0}^n A_i f(x_i).$$

Thus, the coefficients A_i can be computed by integration of the polynomials $p_i(x)$. However, in the sequel we shall use other methods for computing the coefficients, e.g. the "method of unknown coefficients." Here the coefficients are determined by putting

$$\int_a^b f(x)\, dx = \sum_{i=0}^n A_i f(x_i) + R_\mathrm{T},$$

and then requiring that the truncation error R_T be equal to zero for the functions

$$f(x) = x^k, \qquad k = 0, 1, \ldots, n.$$

This condition gives a linear system of equations for the coefficients A_i.

The most commonly used Newton–Cote's formulas are those obtained with $n = 1$, the **trapezoidal rule**, and for $n = 2$, **Simpson's rule**. Let us derive these formulas! In the rest of this chapter, we put $f(x_i) = f_i$.

In the **trapezoidal rule**, f is approximated by a straight line through the points (x_0, f_0) and (x_1, f_1) (see Figure 7.2.1).

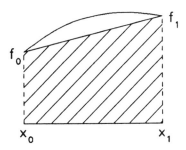

Figure 7.2.1   The trapezoidal rule.

This means that the integral is approximated by the area of a trapezoid, and we immediately get

$$\int_{x_0}^{x_1} f(x)\, dx = \frac{h}{2}(f_0 + f_1) + R_\mathrm{T}.$$

From the argument above, the truncation error is seen to be

$$R_\mathrm{T} = \int_{x_0}^{x_1} (x - x_0)(x - x_1)\frac{f''(\xi(x))}{2!}\, dx,$$

where $\xi(x)$ is a point in the interval (x_0, x_1). To simplify this expression, we make the change of variables $x = x_0 + ph$. Since $x_1 = x_0 + h$, we get

$$R_\mathrm{T} = h\int_0^1 ph(p-1)h\frac{f''(\xi(x_0 + ph))}{2!}\, dp$$

$$= \frac{h^3}{2}\int_0^1 p(p-1)f''(\xi(x_0 + ph))\, dp.$$

This can be simplified further using the mean value theorem of integral calculus:

Theorem 7.2.2   If the function f is continuous, and the function g is continuous and does not change sign in the closed interval $[a, b]$, then there is a point ξ inside the interval, such that

$$\int_a^b f(x)g(x)\, dx = f(\xi) \int_a^b g(x)\, dx.$$

Again study the expression for R_T! The function $p(p-1)$ has constant sign for $0 \le p \le 1$. If f'' is assumed to be continuous in (x_0, x_1), then the mean value theorem gives

$$R_T = \frac{h^3}{2} f''(\xi(x_0 + \bar{p}h)) \int_0^1 p(p-1)\, dp,$$

where $0 < \bar{p} < 1$. We put $\eta = \xi(x_0 + \bar{p}h)$, compute the integral, and get

$$R_T = -\frac{h^3}{12} f''(\eta), \qquad x_0 < \eta < x_1.$$

In **Simpson's rule**, the integrand f is approximated by the second degree polynomial through the points $(x_0, f_0), (x_1, f_1)$ and (x_2, f_2). Let us use the method of undetermined coefficients to derive the formula. We write

$$\int_{x_0}^{x_2} f(x)\, dx = h(A_0 f(x_0) + A_1 f(x_1) + A_2 f(x_2)) + R_T,$$

and determine the coefficients A_0, A_1 and A_2 so that the truncation error is zero for polynomials of as high degree as possible.

The change of variables $x = x_1 + t$ gives

$$\int_{x_0}^{x_2} f(x)\, dx = \int_{-h}^h f(x_1 + t)\, dt = \int_{-h}^h g(t)\, dt,$$

where $g(t) = f(x_1 + t)$. Therefore, we consider instead the formula

$$\int_{-h}^h g(t)\, dt = h(A_0 g(-h) + A_1 g(0) + A_2 g(h)) + R_T.$$

The requirement $R_T = 0$ for $g = 1, t, t^2$ gives

$$2h = h(A_0 + A_1 + A_2),$$
$$0 = h(-A_0 h + A_2 h),$$
$$\frac{2h^3}{3} = h(A_0 h^2 + A_2 h^2).$$

We solve for A_0, A_1 and A_2:

$$A_0 = A_2 = \frac{1}{3}, \qquad A_1 = \frac{4}{3}.$$

It is easy to check that the truncation error happens to be zero also for $g(t) = t^3$, but not for $g(t) = t^4$. It can be shown that the truncation error is $-h^5 f^{(4)}(\eta)/90$.

Returning to the original interval, we thus get

$$\int_{x_0}^{x_2} f(x)\, dx = \frac{h}{3}(f(x_0) + 4f(x_1) + f(x_2)) - \frac{h^5}{90} f^{(4)}(\eta).$$

It is tempting to believe that the approximation of a certain integral will improve if we approximate the integrand by polynomials of higher degree. This is not necessarily true, as is seen in the following example.

Example 7.2.3 Compute approximations I_n to

$$I = \int_{-1}^{1} \frac{1}{1 + 25x^2}\, dx$$

using Newton–Cote's quadrature formulas for $n = 1, 2, \ldots, 8$ (n is the degree of the approximating polynomial). We use tabulated values for the coefficients A_i for $n > 2$ and get the result:

n	I_n
1	0.038462
2	0.679487
3	0.208145
4	0.237400
5	0.230769
6	0.387045
7	0.289899
8	0.150049

The exact value is $I = \frac{1}{5}(\arctan 5 - \arctan(-5)) \approx 0.549360$. We are not surprised to get bad approximations for small values of n. Too few

points in the interval are used. Why we do not get better approximations for larger n is explained by Figure 5.2.4. When we approximate
this particular integrand, Runge's phenomenon occurs, and the interpolating polynomial deviates very much from the integrand between
the interpolation points.

It has been shown that there are continuous functions f such that the approximations I_n of $I = \int_a^b f(x)\,dx$ computed from Newton–Cote's quadrature formulas do not converge to I as $n \to \infty$. Therefore, we approximate
instead f on subintervals of (a, b) by polynomials of low degree (straight
lines or second degree polynomials).

If the trapezoidal rule is used on the subinterval (x_i, x_{i+1}), we get

$$\int_{x_i}^{x_{i+1}} f(x)\,dx = \frac{h}{2}(f_i + f_{i+1}) - \frac{h^3}{12}f''(\eta_i),$$

where $x_i \leq \eta_i \leq x_{i+1}$. The truncation error $-(h^3/12)f''(\eta_i)$ in the subinterval is called **local truncation error**.

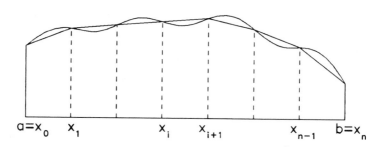

$$a{=}x_0 \quad x_1 \qquad\qquad x_i \quad x_{i+1} \qquad\qquad x_{n-1} \quad b{=}x_n$$

Figure 7.2.4   The trapezoidal rule.

For the whole interval (see Figure 7.2.4), we get

$$\int_a^b f(x)\,dx = \sum_{i=0}^{n-1} \int_{x_i}^{x_{i+1}} f(x)\,dx$$

$$= h\left(\frac{1}{2}f_0 + f_1 + \ldots + f_{n-1} + \frac{1}{2}f_n\right) + R_T.$$

The **total truncation error** R_T is the sum of the local errors, i.e.,

$$R_T = -\frac{h^3}{12}\sum_{i=0}^{n-1} f''(\eta_i).$$

Note that $b - a = nh$, and therefore

$$R_T = -\frac{b-a}{12}h^2 \, \frac{\sum_{i=0}^{n-1} f''(\eta_i)}{n}.$$

If f'' is continuous in (a, b), then there is a number η in this interval, such that

$$\frac{1}{n}\sum_{i=0}^{n-1} f''(\eta_i) = f''(\eta).$$

Hence the total truncation error for the trapezoidal rule is

$$R_T = -\frac{b-a}{12}h^2 f''(\eta).$$

In particular, $\lim_{h\to 0} R_T = 0$ holds. Therefore, the finer the subdivision of the interval, the better the approximation to the integral.

We summarize:

The trapezoidal rule

$$\int_a^b f(x)\,dx = T(h) + R_T,$$

$$T(h) = h\left(\frac{1}{2}f_0 + f_1 + \ldots + f_{n-1} + \frac{1}{2}f_n\right).$$

The truncation error is

$$R_T = -\frac{b-a}{12}h^2 f''(\eta),$$

$$h = \frac{b-a}{n}$$

if f'' is continuous.

Example 7.2.5 Compute $\int_0^1 1/(1+x)\,dx$ using the trapezoidal rule. Choose $h = 1, 0.5, 0.25$ and 0.125, and compare the results to the exact value $\log 2 \approx 0.693147$.

We get

h	$T(h)$	$T(h) - \log 2$
1	0.750000	0.056853
0.5	0.708333	0.015186
0.25	0.697024	0.003877
0.125	0.694122	0.000975

Since $R_T = O(h^2)$, the error $T(h) - \log 2$ should be divided by a factor of four when h is halved. We see that this happens in the example.

If $n = 2m$, i.e., the interval (a, b) is subdivided into an *even* number of subintervals, then the integral in each subinterval (x_{2i}, x_{2i+2}) can be computed from Simpson's rule:

$$\int_{x_{2i}}^{x_{2i+2}} f(x)\, dx = \frac{h}{3}(f_{2i} + 4f_{2i+1} + f_{2i+2}) - \frac{h^5}{90}f^{(4)}(\eta_i),$$

where $x_{2i} \le \eta_i \le x_{2i+2}$. For the whole interval we get

$$\int_a^b f(x)\, dx = \sum_{i=0}^{m-1} \int_{x_{2i}}^{x_{2i+2}} f(x)\, dx$$

$$= \frac{h}{3}(f_0 + 4f_1 + 2f_2 + 4f_3 + \ldots + 4f_{2m-1} + f_{2m}) + R_T,$$

where

$$R_T = -\frac{h^5}{90}\sum_{i=0}^{m-1} f^{(4)}(\eta_i).$$

Using the same argument as earlier for the trapezoidal rule, this can be written

$$R_T = -\frac{b-a}{180}h^4 f^{(4)}(\eta),$$

where $\eta \in (a, b)$, provided that $f^{(4)}$ is continuous in (a, b).

We summarize:

Simpson's rule

$$\int_a^b f(x)\, dx = S(h) + R_T,$$

$$S(h) = \frac{h}{3}(f_0 + 4f_1 + 2f_2 + 4f_3 + \ldots + 2f_{2m-2} + 4f_{2m-1} + f_{2m}).$$

The truncation error is

$$R_T = -\frac{b-a}{180}h^4 f^{(4)}(\eta),$$

if $f^{(4)}$ is continuous.

Example 7.2.6 Use Simpson's rule to compute

$$\int_0^1 \frac{1}{1+x}\,dx.$$

Choose $h = 0.5, 0.25$ and 0.125, and compare the results to the exact value.

We get

h	$S(h)$	$S(h) - \log 2$
0.5	0.694444	0.001297
0.25	0.693254	0.000107
0.125	0.693155	0.000008

Since $R_T = O(h^4)$, the error $S(h) - \log 2$ should be divided by 16 when h is halved. We see that this is the case.

7.3 Romberg's Method

As before, let

$$T(h) = h\left(\frac{1}{2}f_0 + f_1 + f_2 + \ldots + f_{n-1} + \frac{1}{2}f_n\right)$$

be an approximation to

$$I = \int_a^b f(x)\,dx.$$

Here

$$h = \frac{b-a}{n},$$
$$x_i = a + ih, \quad i = 0, 1, \ldots, n$$
$$f(x_i) = f_i.$$

We have shown that the truncation error is $O(h^2)$ if f'' is continuous. One can prove even more:

If f is $(2k+2)$ times continuously differentiable for $a \le x \le b$, then

$$T(h) = \int_a^b f(x)\,dx + a_1 h^2 + a_2 h^4 + \ldots + a_k h^{2k} + O(h^{2k+2}), \quad (7.3.1)$$

where the coefficients $a_1, a_2, \ldots, a_k$ are independent of h.

Therefore, we can compute approximations $T(h)$ to the integral using the trapezoidal rule for different values of h, and then use Richardson extrapolation to reduce the truncation error. This procedure is called **Romberg's method**.

Example 7.3.2 Use the values of $T(h)$ from Example 7.2.5 and Richardson extrapolation to approximate $\int_0^1 1/(1+x)\,dx$.

h	$T(h)$	$\Delta/3$	$\Delta/15$	$\Delta/63$	
1	0.750000				
		−13889			
0.5	0.708333		0.694444		
		−3770		−79	
0.25	0.697024		0.693254		0.693175
		−967		−7		0
0.125	0.694122		0.693155	0.693148	0.693148

(The corrections are given in units of 10^{-6}.)

The exact value $\log 2 \approx 0.693147$ is very accurately approximated.

In the first step of extrapolation, the h^2-term in the truncation error is eliminated, and we get approximations with truncation error $O(h^4)$. Simpson's rule also gives this truncation error. If we compare Example 7.2.6 to Example 7.3.2, we see that one Richardson extrapolation of values computed with the trapezoidal rule gives exactly the same result as Simpson's rule:

$$S(h) = T(h) + \frac{1}{3}(T(h) - T(2h)).$$

In order to show that this is valid in general, we write the sums defining $T(h)$ and $T(2h)$ on top of each other:

$$h(\tfrac{1}{2}f(a) + f(a+h) + f(a+2h) + \ldots + f(b-2h) + f(b-h) + \tfrac{1}{2}f(b)),$$

$$2h(\tfrac{1}{2}f(a) \qquad\qquad + f(a+2h) + \ldots + f(b-2h) \qquad\qquad + \tfrac{1}{2}f(b)).$$

Now it is easy to obtain the extrapolated value

$$
\begin{aligned}
T(h) + \frac{1}{3}(T(h) - T(2h)) &= \frac{1}{3}(4T(h) - T(2h)) \\
&= \frac{h}{3}(f(a) + 4f(a+h) + 2f(a+2h) + \ldots \\
&\quad + 2f(b-2h) + 4f(b-h) + f(b)) = S(h).
\end{aligned}
$$

In repeated Richardson extrapolation we do not get the same result as with integration of higher degree interpolating polynomials. It can be shown that the entries in each column of the Romberg scheme converge to the integral as $h \to 0$.

The truncation error for any value in the scheme is estimated, as before, as the difference between that value and the nearest value above in the same column.

Also, the errors in the function values that are used in the computation of $T(h)$ give rise to errors. Assume that the function values have absolute errors no larger than ϵ in magnitude. For the values in the first column, we have

$$|\bar{T}(h) - T(h)| \le h(\frac{1}{2}|\bar{f}_0 - f_0| + |\bar{f}_1 - f_1| + \ldots + \frac{1}{2}|\bar{f}_n - f_n|)$$

$$\le h(\frac{1}{2}\epsilon + (n-1)\epsilon + \frac{1}{2}\epsilon) = hn\epsilon$$

$$= (b-a)\epsilon.$$

It can be shown that the same is valid for all columns in the Romberg scheme.

If the function values have absolute errors no larger than ϵ in magnitude, then these lead to an error R_{XF} in any entry of the Romberg scheme that can be estimated as

$$|R_{\mathrm{XF}}| \le (b-a)\epsilon.$$

Example 7.3.3 We estimate the total error in the approximate value 0.693148 of $\int_0^1 1/(1+x)\,dx$ that was computed in Example 7.3.2. For the truncation error, we have

$$|R_{\mathrm{T}}| \le |0.693148 - 0.693175| \le 2.7 \cdot 10^{-5}.$$

The function values have absolute errors of at most $0.5 \cdot 10^{-6}$. These errors give an error in the result that can be estimated as

$$|R_{\mathrm{XF}}| \le 1 \cdot 0.5 \cdot 10^{-6}.$$

If we disregard the error in the computation of the corrections, then the total error is $3 \cdot 10^{-5}$, at most.

In the computation of the values of the first column of the Romberg scheme, one can use the previously computed trapezoidal rule approximations. This is illustrated in a part of a program, where $T_i = T\big((b-a)/2^{i-1}\big)$ are computed for $i = 1, 2, \ldots,$ rowmax.

$$h := b - a;$$
$$T_1 := h * (f(a) + f(b))/2;$$
for $i := 2$ **to** rowmax **do**
    ($*$ function values in new points are summed up $*$)
    $h := h/2;$
    $T_i := 0;$
    $j := 1;$
    while $(a + (2j - 1)h < b)$ **do**
        $T_i := T_i + f(a + (2j - 1)h);$
        $j := j + 1$
    $T_i := h * T_i + T_{i-1}/2;$
    ($*T_{i-1}$ is divided by 2 since $*$)
    ($*$ the step length has been halved. $*$)

7.4 Difficulties in Numerical Integration

To be able to compute the integral of a function over a finite interval using Romberg's method, $T(h)$ must have an expansion of the type (7.3.1). This is not always the case. For instance, the integrand or some of its derivatives may have an endpoint singularity, i.e., be of the form $f(x) = (x - a)^\alpha g(x)$ or $f(x) = (b - x)^\alpha g(x)$, where $\alpha \neq 0$, $-1 < \alpha < 1$. If the function g is $(2k + 1)$ times continuously differentiable in the interval of integration, and, for negative α, $f(a) = 0$ or $f(b) = 0$, then the truncation error can be shown to be

$$R_T = \sum_{i=1}^{k} a_i h^{2i} + \sum_{1 \le i \le 2k+1-\alpha} b_i h^{i+\alpha} + O(h^{2k+1}).$$

In Richardson extrapolation, we must then also take into account the terms $b_1 h^{1+\alpha}$, $b_2 h^{2+\alpha}$, etc.

In many cases, however, the integral can be rewritten into a form that is better suited for numerical computation. It is a good rule to use numerical integration only when the integral has been simplified as far as possible by mathematical reductions.

We shall use the integral

$$I = \int_0^1 \frac{\arctan x}{x^{3/2}} \, dx$$

to demonstrate different ways to handle an endpoint singularity.
For $|x| \le 1$, we have

$$\arctan x = x - \frac{x^3}{3} + \frac{x^5}{5} - \ldots = x + O(x^3).$$

Hence the integrand has a singularity for $x = 0$:

$$\frac{\arctan x}{x^{3/2}} = x^{-1/2} + O(x^{3/2}).$$

Partial integration gives

$$I = \int_0^1 \frac{\arctan x}{x^{3/2}} \, dx = \left[-2x^{-1/2} \arctan x \right]_0^1 + \int_0^1 \frac{2}{x^{1/2}(1 + x^2)} \, dx.$$

With the change of variables $t = x^{1/2}$, we get

$$I = -\frac{\pi}{2} + \int_0^1 \frac{4}{1 + t^4} \, dt.$$

The latter integral can be computed numerically. It is even possible to give
an exact expression for the primitive function here.
We could have made a **change of variables** directly:

$$I = \int_0^1 \frac{\arctan x}{x^{3/2}} \, dx = 2 \int_0^1 \frac{\arctan t^2}{t^2} \, dt.$$

The integrand has the limit 1 at the origin, and the integral can be computed numerically.
"Subtraction of the singularity" means that a function with the
same type of singularity and with a known primitive function is subtracted
from the integrand. In our example we get

$$I = \int_0^1 \frac{\arctan x - x}{x^{3/2}} \, dx + \int_0^1 \frac{1}{\sqrt{x}} \, dx = \int_0^1 \frac{\arctan x - x}{x^{3/2}} \, dx + 2.$$

The new integrand is of the form $-\frac{1}{3}x^{3/2} + O(x^{7/2})$ close to the origin.
This means that its second derivative is singular at the origin and that
Romberg's method can be expected to give bad accuracy. For small values
of x, we also have cancellation in the computation of the integrand.
Series expansion and termwise integration often work well. In our
case, we get

$$I = \int_0^1 \sum_{n=0}^\infty (-1)^n \frac{x^{2n+1}}{2n + 1} x^{-3/2} \, dx = \sum_{n=0}^\infty (-1)^n \frac{2}{(2n + 1)(4n + 1)}.$$

The series to the right converges so slowly that in this case it is better to use another method.

Another type of difficulty occurs if the interval of integration is infinite. Sometimes a change of variables can be used to transform the interval into a finite one. E.g., the interval $(0, \infty)$ is transformed into $(1, 0)$ by $t = 1/(x+1)$ or $t = e^{-x}$.

In some cases an infinite interval of integration can be approximated by a finite interval. If, e.g., $\int_0^\infty e^{-x^2}\, dx$ is approximated by $\int_0^a e^{-x^2}\, dx$, we make a truncation error

$$R_{\mathrm{T}} = \int_a^\infty e^{-x^2}\, dx < \frac{1}{a}\int_a^\infty xe^{-x^2}\, dx = \frac{1}{2a}e^{-a^2}.$$

For $a = 5$, we have $R_{\mathrm{T}} < 1.4 \cdot 10^{-12}$.

A third difficulty occurs if the integrand is oscillating heavily. Special methods have been developed for this case.

7.5   Adaptive Quadrature

In many cases, our goal is to approximate a given integral to a certain accuracy using as few function evaluations as possible. If the integral varies rapidly in part of the interval of integration, then a small step length is needed to get a satisfactory accuracy in this subinterval, while in the rest of the interval it would be possible to get the same accuracy with a much larger step. To avoid making unnecessarily many function evaluations, we must adjust the step length to the behaviour of the integrand. This is done in adaptive quadrature.

For instance, we can use the following principle: approximations I_1 and I_2 of the integral are computed using Simpson's rule with step lengths $(b - a)/2$ and $(b - a)/4$. Then I_3 is computed using extrapolation. If $|I_3 - I_2|/|I_2| < \epsilon$, where ϵ is a given tolerance, then I_3 is accepted as an approximation of the integral. Otherwise the interval (a, b) is partitioned in two parts (a, m) and (m, b), where $m = (a + b)/2$. The integrals over these subintervals are then computed independently. With this procedure, a fine partitioning is made only where it is needed to obtain the required accuracy. Using a recursive function, this can be programmed very easily.

Consider, as an example,

$$\int_0^4 \frac{32}{1 + 1024x^2}\, dx.$$

The integrand has a maximum at the origin and decreases very rapidly to zero. Using a program for adaptive quadrature, working according to the principles outlined above, 33 function evaluations were needed to compute the integral with three correct decimals. The table below shows the smallest step length that was used in successive subintervals.

Interval		Step Length
0	0.0625	1/128
0.0625	0.125	1/64
0.125	0.25	1/32
0.25	0.5	1/16
0.5	1	1/8
1	2	1/4
2	4	1/2

We see that very small step lengths were needed close to the origin. To compute the integral with three correct decimals using Romberg's method, 257 evaluations of the integrand were needed (in fact, the best value was obtained with the trapezoidal rule for $h = 1/64$, without extrapolation.)

Exercises

1. Compute $\int_0^1 t^3 \, dt$ using Romberg's method. Use the step lengths $h = 1$ and $h = \frac{1}{2}$. Explain why the result agrees with the exact value.

2. A car is started in cold weather, and a flux meter is used to measure the consumption of gasoline. The following values are obtained.

x	0	1.25	2.5	3.75	5	6.25	7.5	8.75	10
$f(x)$	0.260	0.208	0.172	0.145	0.126	0.113	0.104	0.097	0.092

(x is the distance (km), $f(x)$ is the instantaneous flux (liters/km).) The values of $f(x)$ are assumed to be correctly rounded. Compute the gas consumption for 10 km with an absolute error smaller than 10^{-2}.

3. Consider the integral

$$F(a) = \int_0^1 e^{-a^2 x^2} \, dx.$$

a) Compute $F(1)$ with an absolute error not larger than 10^{-5}.

b) Assume that $a = 1 \pm \epsilon$, where $\epsilon \ll 1$. Give a bound for the error that arises because we use the approximation $\bar{a} = 1$ for the computation of $F(a)$.

4. Compute

$$I = \int_0^4 \frac{dx}{1 + 5xe^{x^2}}$$

with two correct decimals. Sketch the integrand, and use Romberg's method on separate subintervals.

5. To compute the integral

$$\int_0^1 \frac{\sqrt{x}}{4 - x^2}\, dx,$$

we used a program for Romberg's method based on formula (7.3.1). The following results were obtained.

h	$T(h)$			
0.5	0.17761420			
0.25	0.18353680	0.18551100		
0.125	0.18622450	0.18712040	0.18722770	
0.0625	0.18732990	0.18769830	0.18773690	0.18774500

a) Explain the slow convergence.
b) Compute a better approximation to the integral using the values in the table.

6. Suggest different methods for the numerical computation of

$$\int_0^1 \frac{\sin x}{\sqrt{1 - x^2}}\, dx.$$

7. Suggest a method for the numerical computation of

$$\int_0^\infty \frac{e^{-x}}{1 + xe^{-x}}\, dx.$$

References

For a more extensive treatment of extrapolation methods for numerical integration, we refer to

J. Stoer and R. Bulirsch, *Introduction to Numerical Analysis*, Springer Verlag, 1983.

A recursive Pascal program for adaptive quadrature is given in

W. Gander, *Computermathematik*, Birkhäuser Verlag, 1985.

8   Systems of Linear Equations

8.1   Introduction

Linear systems of equations are very common in scientific computations. When a boundary value problem for an ordinary or a partial differential equation is discretized, it is reduced to a linear system of equations (see Chapters 1 and 10). Such problems arise, e.g., in structural mechanics and fluid dynamics, and the number of unknowns can be very large, sometimes of the order of 10^5 or more.

Since linear systems of equations are so common, it is important to be able to solve them efficiently and accurately. During the last decades, high quality software for the computer solution of linear systems has been developed, and much research is presently being done concerning the solution of large systems on vector and parallel computers. To be able to use such software, it is essential to know the properties of the basic algorithms.

In this chapter, we shall study the numerical solution of linear systems of equations with n unknowns

$$a_{11}x_1 + a_{12}x_2 + \ldots + a_{1n}x_n = b_1,$$
$$a_{21}x_1 + a_{22}x_2 + \ldots + a_{2n}x_n = b_2,$$
$$\vdots$$
$$a_{n1}x_1 + a_{n2}x_2 + \ldots + a_{nn}x_n = b_n.$$

All the coefficients are assumed to be real. Often we write the system in matrix form

$$Ax = b,$$

where A is a $n \times n$ matrix, and b an $n \times 1$ vector:

$$A = \begin{pmatrix} a_{11} & a_{12} & \cdots & a_{1n} \\ a_{21} & a_{22} & \cdots & a_{2n} \\ \vdots & \vdots & & \vdots \\ a_{n1} & a_{n2} & \cdots & a_{nn} \end{pmatrix}, \qquad b = \begin{pmatrix} b_1 \\ b_2 \\ \vdots \\ b_n \end{pmatrix}.$$

Sometimes it is useful to regard the matrix as n column vectors:

$$A = [a_{\cdot 1} a_{\cdot 2} \ldots a_{\cdot n}], \qquad a_{\cdot j} = \begin{pmatrix} a_{1j} \\ a_{2j} \\ \vdots \\ a_{nj} \end{pmatrix}.$$

Then we can write the system $Ax = b$ in the form

$$x_1 a_{\cdot 1} + x_2 a_{\cdot 2} + \ldots + x_n a_{\cdot n} = b.$$

The problem of solving the system of equations is therefore equivalent to representing b as a linear combination of the vectors $a_{\cdot 1}, a_{\cdot 2}, \ldots a_{\cdot n}$. We see that the system has a solution for any given right hand side b if and only if the column vectors constitute a **basis** in R^n, or, equivalently, the column vectors are **linearly independent**. A matrix A, whose column vectors are linearly independent, is said to be **nonsingular**. If A is nonsingular, then $Ax = b$ has a unique solution for any right hand side b. Unless otherwise stated, we assume in this chapter that the linear systems of equations that we are considering have nonsingular matrices.

There are two main classes of methods for the numerical solution of linear systems of equations, **direct** and **iterative methods**. With a direct method, the solution is computed by performing a finite number of arithmetic operations. If it were possible to do this without rounding errors, then the computed solution would be the exact solution. With an iterative method, a sequence of vectors $x^{(0)}, x^{(1)}, x^{(2)}, \ldots$ is formed, which converges to the solution of $Ax = b$.

As a rule, direct methods are used when most of the coefficients of the matrix A are nonzero, while iterative methods are preferred for **sparse** systems, i.e., systems where a majority of the coefficients are equal to zero (the latter type of linear system typically occurs when a differential equation is discretized, but also in many other contexts). In this book, we only treat direct methods.

The most basic direct method for solving a linear system of equations is **Gaussian elimination**. In this method, multiples of the equations are added to each other in a systematic way so that the system becomes triangular. Using matrix formulation, this can be written

$$[A \; \vdots \; b] \rightarrow [U \; \vdots \; c],$$

where U is upper triangular. The solution is then obtained by back substitution in the system

$$Ux = c.$$

We start by studying the second step in the solution procedure, the solution of triangular systems.

8.2 Triangular Systems

An $n \times n$ matrix $U = (u_{ij})$ is said to be **upper triangular** if

$$u_{ij} = 0, \qquad i > j,$$

and an $n \times n$ matrix $L = (l_{ij})$ is said to be **lower triangular** if

$$l_{ij} = 0, \qquad i < j.$$

Here, we shall only discuss the solution of upper triangular systems $Ux = c$, since lower triangular systems can be solved analogously.

Provided that all the diagonal elements of the matrix U are nonzero (U is nonsingular), the system

$$
\begin{pmatrix}
u_{11} & u_{12} & \cdots & u_{1n} \\
 & u_{22} & \cdots & u_{2n} \\
 & & & \vdots \\
 & & & u_{nn}
\end{pmatrix}
\begin{pmatrix}
x_1 \\ x_2 \\ \vdots \\ x_n
\end{pmatrix}
=
\begin{pmatrix}
c_1 \\ c_2 \\ \vdots \\ c_n
\end{pmatrix},
$$

can be solved by **back substitution**:

$$x_n = c_n / u_{nn},$$

$$x_i = \left(c_i - \sum_{j=i+1}^{n} u_{ij} x_j \right) / u_{ii}, \qquad i = n-1, n-2, \ldots, 1.$$

It is important to know approximately how much work (computer time) is needed to solve a system of equations. One measure of the amount of work is the number of arithmetic operations that are performed. From the formulas above, we see that the computation of x_i requires $n - i$ additions and multiplications (and one division). We let "one operation" denote "one addition and one multiplication". The number of operations for solving a triangular system with n unknowns will then be

$$\sum_{i=1}^{n-1}(n-i) = \sum_{k=1}^{n-1} k = \frac{n(n-1)}{2} = \frac{n^2 - n}{2}.$$

For large values of n, the n^2 term dominates (note that the operation count is not relevant as a measure of the amount of work for solving a *small* system on a computer, since in this situation other operations, e.g., operative system overhead, will take longer time than actually performing the arithmetic operations). Further, we see that since the number of divisions is n, these operations are less important when we estimate the total amount of work (on modern computers a division does not take much longer to perform than a multiplication).

Thus, we get the following rule of thumb.

> Approximately $n^2/2$ operations are required to solve a triangular system of equations with n unknowns.

8.3   Gaussian Elimination

We start by considering a small system $Ax = b$, where A is a 3×3 matrix,

$$\begin{pmatrix} 5 & -5 & 10 \\ 2 & 0 & 8 \\ 1 & 1 & 5 \end{pmatrix} \begin{pmatrix} x_1 \\ x_2 \\ x_3 \end{pmatrix} = \begin{pmatrix} -25 \\ 6 \\ 9 \end{pmatrix}. \qquad \begin{matrix} (1) \\ (2) \\ (3) \end{matrix}$$

The goal in Gaussian elimination is to reduce the matrix of coefficients to upper triangular form. This is done as follows: First x_1 is eliminated from Equations (2) and (3) by subtracting multiples of Equation (1). (We do not write the vector of unknowns explicitly.)

$$\begin{pmatrix} 5 & -5 & 10 & \vdots & -25 \\ 0 & 2 & 4 & \vdots & 16 \\ 0 & 2 & 3 & \vdots & 14 \end{pmatrix}. \qquad \begin{matrix} (1) \\ (2') = (2) - \frac{2}{5} \cdot (1) \\ (3') = (3) - \frac{1}{5} \cdot (1) \end{matrix}$$

The operations that we have performed can also be described as follows: the elements in positions (2,1) (i.e., row 2, column 1) and (3,1) are zeroed by subtracting multiples of row 1. In the sequel, we shall use this terminology.

In the next step, we shall zero the element in position (3,2) by subtracting a multiple of row 2:

$$\begin{pmatrix} 5 & -5 & 10 & \vdots & -25 \\ 0 & 2 & 4 & \vdots & 16 \\ 0 & 0 & -1 & \vdots & -2 \end{pmatrix}. \qquad \begin{matrix} (1) \\ (2') \\ (3'') = (3') - 1 \cdot (2') \end{matrix}$$

Back substitution gives

$$\begin{cases} x_3 = 2, \\ x_2 = 4, \\ x_1 = -5. \end{cases}$$

We now describe Gaussian elimination applied to a system with n unknowns:

$$[A \vdots b] = \begin{pmatrix} a_{11} & a_{12} & \cdots & a_{1n} & \vdots & b_1 \\ a_{21} & a_{22} & \cdots & a_{2n} & \vdots & b_2 \\ \vdots & \vdots & & \vdots & \vdots & \vdots \\ a_{n1} & a_{n2} & \cdots & a_{nn} & \vdots & b_n \end{pmatrix}.$$

In the first step, we zero elements in the first column by subtracting multiples of row 1 (we presuppose that $a_{11} \neq 0$). The result of this step is

$$\begin{pmatrix} a_{11} & a_{12} & \cdots & a_{1n} & \vdots & b_1 \\ 0 & a'_{22} & \cdots & a'_{2n} & \vdots & b'_2 \\ \vdots & \vdots & & \vdots & \vdots & \vdots \\ 0 & a'_{n2} & \cdots & a'_{nn} & \vdots & b'_n \end{pmatrix}.$$

Matrix elements that have been changed in this transformation are marked with a prime. Here, we do not describe how the matrix elements have been changed, but go into details later.

In the next step, we zero elements in column 2 below the main diagonal (below position (2,2)) by subtracting multiples of row 2 (provided that $a'_{22} \neq 0$).

After $k - 1$ steps the matrix has been transformed into

$$\begin{pmatrix} a_{11} & a_{12} & & \cdots & & a_{1n} & \vdots & b_1 \\ & a_{22} & & \cdots & & a_{2n} & \vdots & b_2 \\ & & \ddots & & & \vdots & \vdots & \vdots \\ & & & a_{kk} & a_{k,k+1} & \cdots & a_{kn} & \vdots & b_k \\ & & & \vdots & \vdots & & \vdots & \vdots & \vdots \\ & & & a_{ik} & a_{i,k+1} & \cdots & a_{in} & \vdots & b_i \\ & & & \vdots & \vdots & & \vdots & \vdots & \vdots \\ & & & a_{nk} & a_{n,k+1} & \cdots & a_{nn} & \vdots & b_n \end{pmatrix}.$$

Note that, for simplicity, we do not use a special notation to denote that the elements have been changed. Thus a_{ij} in this matrix has a different value than in the original matrix.

We assume that $a_{kk} \neq 0$. Now, we zero the elements in column k under the main diagonal by subtracting multiples of row k. The result is

$$
\begin{pmatrix}
a_{11} & a_{12} & & \cdots & & a_{1n} & \vdots & b_1 \\
 & a_{22} & & \cdots & & a_{2n} & \vdots & b_{2n} \\
 & & \ddots & & & \vdots & \vdots & \vdots \\
 & & a_{kk} & a_{k,k+1} & \cdots & a_{kn} & \vdots & b_k \\
 & & \vdots & \vdots & & \vdots & \vdots & \vdots \\
 & & 0 & a'_{i,k+1} & \cdots & a'_{in} & \vdots & b'_i \\
 & & \vdots & \vdots & & \vdots & \vdots & \vdots \\
 & & 0 & a'_{n,k+1} & \cdots & a'_{nn} & \vdots & b'_n
\end{pmatrix},
$$

where the transformed elements are given by

$$ a'_{ij} = a_{ij} - m_{ik}a_{kj}, \qquad j = k+1,\ldots,n, \quad i = k+1,\ldots,n, $$

$$ b'_i = b_i - m_{ik}b_k, \qquad i = k+1,\ldots,n, $$

with

$$ m_{ik} = \frac{a_{ik}}{a_{kk}}, \qquad i = k+1,\ldots,n. $$

Row k, which is used to zero the elements in column k in this transformation, is called the **pivot row**, and a_{kk} is called the **pivot element**. To confirm that this transformation zeroes the elements in column k, we compute

$$ a'_{ik} = a_{ik} - m_{ik}a_{kk} = a_{ik} - \frac{a_{ik}}{a_{kk}}a_{kk} = 0. $$

We formulate the transformations of this step in Pascal-like code:

```
for i := k + 1 to n do m_ik := a_ik/a_kk;
for all (i, j), k + 1 ≤ i, j ≤ n do
    a_ij := a_ij − m_ik * a_kj;
for i := k + 1 to n do b_i := b_i − m_ik * b_k;
```

After $n-1$ steps of this algorithm, we have reduced the matrix of coefficients to upper triangular form. All the steps are similar to step k, and therefore we get an algorithm for Gaussian elimination (provided that all the pivot elements are nonzero, and with certain other provisos, see Section 8.4) if we repeat the above code for $k = 1,\ldots,n-1$:

```
for k := 1 to n − 1 do
    for i := k + 1 to n do m_{ik} := a_{ik}/a_{kk};
    for all (i, j), k + 1 ≤ i, j ≤ n do
        a_{ij} := a_{ij} − m_{ik} * a_{kj};
    for i := k + 1 to n do b_i := b_i − m_{ik} * b_k;
```

We shall now estimate how much work is needed to perform Gaussian elimination. As before, we let "one operation" denote "one addition and one multiplication." Since we want to estimate the dominating term, we only consider the transformations of the matrix elements and disregard the divisions and the operations performed to the right hand side.

We consider step k in Gaussian elimination, and we see that, in this step, $(n-k)^2$ matrix elements are transformed, and that one operation is performed on each element. Since we have $(n-1)$ steps, and k is repeated from 1 to $n-1$, the total amount of operations is seen to be

$$\sum_{k=1}^{n-1}(n-k)^2 = \sum_{\nu=1}^{n-1}\nu^2.$$

The following lemma holds.

Lemma 8.3.1

$$\sum_{\nu=1}^{n-1}\nu^2 = \frac{n^3}{3} - \frac{n^2}{2} + \frac{n}{6}.$$

Proof. (Induction over n.) The formula is obviously true for $n = 2$. Assume that it holds for $n = N$, and put

$$S_N = \frac{N^3}{3} - \frac{N^2}{2} + \frac{N}{6}.$$

We then have

$$S_{N+1} = \frac{(N+1)^3}{3} - \frac{(N+1)^2}{2} + \frac{N+1}{6},$$

and after some elementary reformulations we get

$$S_{N+1} = S_N + N^2,$$

which proves the lemma. ∎

For large n, the n^3 term dominates, and thus we have shown the following rule of thumb.

> The reduction of a linear system of equations with n unknowns to triangular form (Gaussian elimination) requires approximately $n^3/3$ operations.

Assume that an operation takes $5\mu s$ on a certain computer. The following table gives the times, for different values of n, to perform Gaussian elimination and back substitution on this computer. The times are based on our rules of thumb.

n	Elimination	Back Substitution
100	1.7	0.025
1000	$1.7 \cdot 10^3 (\approx 28 \text{ min})$	2.5
10000	$1.7 \cdot 10^6 (\approx 20 \text{ days})$	250

Table 8.3.2   Time in seconds for Gaussian elimination and back substitution. One operation is assumed to take $5\mu s$.

Table 8.3.2 shows that linear systems with a few hundred unknowns can easily be solved on modern computers.

The main use of operation counts is not to predict accurately how long it takes to solve a system of a certain dimension on a given computer. As a matter of fact, this is impossible, since one must also take into account operative system overhead. Further, in a time-sharing environment several programs are executed simultaneously. However, operation counts can be used to compare different algorithms for solving the same problem, and to estimate the cost (in terms of CPU time).

Example 8.3.3   The solution of a linear system with n unknowns takes 2 CPU seconds on a certain computer. Approximately how much CPU time is required to solve a system with $3n$ unknowns?

Since the time required is proportional to the dimension raised to the third power, we can write

$$t = t(n) = C \cdot n^3;$$

we now get

$$t(3n) = C \cdot (3n)^3 = 27Cn^3 = 27t(2) = 27 \cdot 2 = 54.$$

Approximately one CPU minute is needed for solving the system with $3n$ unknowns.

The operation count is based on the assumption that most of the elements of the matrix are nonzero. In many applications, systems arise where a large part (say 90% or more) of the matrix elements are equal to zero. If it is possible to take advantage of this fact, so that elements that are already zero are not zeroed in the elimination, then the operation count can be reduced by one or two orders of magnitude. We shall discuss one such class of matrices, **band matrices**, in Section 8.8.

The use of operation counts as a single measure of the complexity of an algorithm presupposes that the floating point operations constitute the main part of the work in solving a linear system. However, this is not true on many modern computers, where memory accesses may be equally or more important. This is further discussed in Section 8.14.

For the presentation in this section, we have assumed that all the pivot element are nonzero. This is of course not a realistic assumption. The algorithm can easily be modified to handle the case when a zero occurs in the pivot element position. This and other numerical problems are discussed in the next section.

8.4   Pivoting

Now we will consider the situation when a pivot element in Gaussian elimination is equal to zero. Assume that we shall solve the following system

$$\begin{pmatrix} 0 & 1 & \vdots & 1 \\ 1 & 1 & \vdots & 2 \end{pmatrix}.$$

The algorithm from the preceding section must be modified, so that first rows 1 and 2 in the matrix are interchanged:

$$\begin{pmatrix} 1 & 1 & \vdots & 2 \\ 0 & 1 & \vdots & 1 \end{pmatrix}.$$

It is necessary also to interchange rows when the following system is solved:

$$\begin{pmatrix} \epsilon & 1 & \vdots & 1 \\ 1 & 1 & \vdots & 2 \end{pmatrix},$$

for ϵ small, since loss of accuracy can occur due to rounding errors. As an example, let $\epsilon = 10^{-5}$. The multiplier for zeroing in position (2,1) is $m_{21} = 10^5$, and the result becomes

$$\begin{pmatrix} 10^{-5} & 1 & \vdots & 1 \\ 0 & 1-10^5 & \vdots & 2-10^5 \end{pmatrix}.$$

Now, if we use the floating point system $(10, 3, -9, 9)$ (base 10, 3 digits in the fraction, cf. Chapter 2), we get

$$\begin{aligned} 1 - 10^5 &= (0.00001 - 1.000)10^5 \\ &= -0.99999 \cdot 10^5 = -9.9999 \cdot 10^4 \doteq -10.000 \cdot 10^4 \\ &= -1.000 \cdot 10^5 \end{aligned}$$

($\doteq$ is to be read "is rounded to"). In the floating point system, we get

$$\begin{pmatrix} 10^{-5} & 1 & \vdots & 1 \\ 0 & -10^5 & \vdots & -10^5 \end{pmatrix},$$

since $2 - 10^5 \doteq -10^5$ also. This reduced system has the solution

$$\begin{cases} x_1 = 0, \\ x_2 = 1. \end{cases}$$

If, instead, we interchange rows and then eliminate, we get $m_{21} = 10^{-5}$, and

$$\begin{pmatrix} 1 & 1 & \vdots & 2 \\ 0 & 1-10^{-5} & \vdots & 1-2\cdot 10^{-5} \end{pmatrix},$$

which is rounded to

$$\begin{pmatrix} 1 & 1 & \vdots & 2 \\ 0 & 1 & \vdots & 1 \end{pmatrix}.$$

The solution is

$$\begin{cases} x_1 = 1, \\ x_2 = 1. \end{cases}$$

To illustrate the same phenomenon, but in a more realistic setting, we have solved the system on a computer that implements the IEEE floating point standard (a PC with an 8087 floating point processor). Thus, we have used the floating point system $(2, 23, -126, 127)$ (single precision). The results are given in Table 8.4.1.

ϵ	Without Interchange		With Interchange	
10^{-5}	1.0013580	0.9999900	1.0000100	0.9999900
10^{-6}	1.0132790	0.9999990	1.0000010	0.9999990
10^{-7}	1.1920930	0.9999999	1.0000000	0.9999999
10^{-8}	0.0000000	1.0000000	1.0000000	1.0000000

Table 8.4.1 Solution of the system of equations for different values of ϵ, with and without row interchange.

It is obvious that the results with row interchange are as good as can be expected (check by computing the exact solution in each case), while the results without interchange are bad, and, for small ϵ, they are disastrously bad.

In this simple example, we can see clearly the cause of the loss of accuracy. In the floating point computation of $1 - 10^5$ and $2 - 10^5$, the information in the equation $x_1 + x_2 = 2$ just disappears; the subtraction $a - 10^5$ gives exactly the same result for all a in the interval $(-5, 5)$. When very large elements appear during the elimination, some (or, in the worst case, all) of the information in the original coefficients is destroyed. But, when we made a row interchange in the matrix in the example, then the elements of the reduced matrix are of the same magnitude as in the given matrix, and loss of accuracy does not occur.

The procedure of performing row interchanges in Gaussian elimination is called **pivoting**. To avoid loss of accuracy, row interchanges are done systematically, to make the pivot element as large as possible. Consider step k in the elimination:

$$
\begin{pmatrix}
a_{11} & a_{12} & & \cdots & & a_{1n} \\
& a_{22} & & & & a_{2n} \\
& & \ddots & & & \vdots \\
& & a_{kk} & a_{k,k+1} & \cdots & a_{kn} \\
& & \vdots & \vdots & & \vdots \\
& & a_{ik} & a_{i,k+1} & \cdots & a_{in} \\
& & \vdots & \vdots & & \vdots \\
& & a_{nk} & a_{n,k+1} & \cdots & a_{nn}
\end{pmatrix}.
$$

Column k is searched (from the element a_{kk} and downwards) for the largest element in magnitude; more precisely, the row index ν is determined so that

$$|a_{\nu k}| = \max_{k \leq i \leq n} |a_{ik}|.$$

Then rows ν and k are interchanged, and the elimination goes on. This procedure is called **partial pivoting**. (In complete pivoting, one determines the element of largest magnitude in the whole submatrix

$$\begin{pmatrix} a_{kk} & \cdots & a_{kn} \\ \vdots & & \vdots \\ a_{nk} & \cdots & a_{nn} \end{pmatrix},$$

and then both rows and columns are interchanged, so that the largest element is moved to position (k, k). This method is seldom used nowadays.)

We now modify the code for Gaussian elimination so that pivoting is performed. Let indmax be a function that in the assignment

$$\nu := \mathrm{indmax}(a, k, n);$$

gives the value ν, so that

$$|a_{\nu k}| = \max_{k \leq i \leq n} |a_{ik}|.$$

Further let swap be a procedure that in the statement

$$\mathrm{swap}(a, b, k, \nu, n)$$

interchanges rows k and ν (and the corresponding elements of the right hand side). The code is now

---

GAUSSIAN ELIMINATION WITH PARTIAL PIVOTING

$\quad$ **for** $k := 1$ **to** $n - 1$ **do**
$\quad\quad \nu := \mathrm{indmax}(a, k, n);$
$\quad\quad \mathrm{swap}(a, b, k, \nu, n);$
$\quad\quad$ **for** $i := k + 1$ **to** n **do** $m_{ik} := a_{ik}/a_{kk};$
$\quad\quad$ **for all** $(i, j),\ k + 1 \leq i, j \leq n$ **do**
$\quad\quad\quad a_{ij} := a_{ij} - m_{ik} * a_{kj};$
$\quad\quad$ **for** $i := k + 1$ **to** n **do** $b_i := b_i - m_{ik} * b_k;$

The aim of pivoting is to avoid large matrix elements during the elimination, and the loss of accuracy that may occur in operations with large elements. Note that in partial pivoting, all the multipliers m_{ik} are smaller than 1 in magnitude:

$$|m_{ik}| = \left| \frac{a_{ik}}{a_{kk}} \right| \leq 1.$$

When linear systems of equations are solved by Gaussian elimination, one should always do partial pivoting in order to avoid loss of accuracy.

There are two special cases, however, where pivoting is not needed: when the matrix A is

- symmetric and positive definite,
- diagonally dominant.

A symmetric matrix A is said to be **positive definite** if

$$x^T A x > 0$$

for all vectors $x \neq 0$. A matrix A is said to be **diagonally dominant** if

$$\sum_{j=1, j \neq i}^{n} |a_{ij}| \leq |a_{ii}|, \qquad i = 1, 2, \ldots, n,$$

with strict inequality for at least one i.

Both of these classes of matrices occur in important applications, e.g., in the discretizations of certain boundary value problems for differential equations. It can be shown that for these classes no growth of the matrix elements can occur: The element largest in magnitude in A is larger than or equal to all elements that occur in the matrix during the elimination, and therefore no unnecessary loss of accuracy will take place. (This does not mean that all multipliers are smaller than one in magnitude, however.) We discuss positive definite matrices further in Section 8.7.

We illustrate Gaussian elimination with partial pivoting in a small example that we shall also use later.

Example 8.4.2 We want to solve the system of equations $Ax = b$, where

$$A = \begin{pmatrix} 0.6 & 1.52 & 3.5 \\ 2 & 4 & 1 \\ 1 & 2.8 & 1 \end{pmatrix}.$$

In the first step of Gaussian elimination, we shall interchange rows 1 and 2:

$$\begin{pmatrix} 2 & 4 & 1 \\ 0.6 & 1.52 & 3.5 \\ 1 & 2.8 & 1 \end{pmatrix}.$$

The multipliers are $m_{21} = 0.3$ and $m_{31} = 0.5$, and, after the first step, we have

$$\begin{pmatrix} 2 & 4 & 1 \\ 0 & 0.32 & 3.2 \\ 0 & 0.8 & 0.5 \end{pmatrix}.$$

Now, we must pivot again, since the largest element in column 2 (in and under the diagonal) is in the third row:

$$\begin{pmatrix} 2 & 4 & 1 \\ 0 & 0.8 & 0.5 \\ 0 & 0.32 & 3.2 \end{pmatrix}.$$

The multiplier is $m_{32} = 0.4$, and the final result is

$$\begin{pmatrix} 2 & 4 & 1 \\ 0 & 0.8 & 0.5 \\ 0 & 0 & 3 \end{pmatrix}.$$

The reader may have noticed that we did not treat the right hand side b in the elimination in the example. Actually, that is not necessary, since we save the multipliers and information about the pivoting. In Section 8.6, we shall show that Gaussian elimination is equivalent to factoring the matrix A into a product of two triangular matrices.

In the introduction of this chapter, we assumed that the linear system of equations has a unique solution, and noted that this is equivalent to the column vectors of A forming a basis (in R^n, in the case when the matrix is $n \times n$). Now we must investigate whether Gaussian elimination with partial pivoting always works under this assumption (in the beginning of this section, we saw that Gaussian elimination *without* pivoting does not always work).

From the description of the algorithm, we see that it will fail if, in some step k, we cannot find any nonzero pivot element, i.e., if

$$a_{ik} = 0, \qquad i = k, k+1, \ldots, n.$$

But this can only happen if column k is a linear combination of the columns $1, 2, \ldots, k-1$. To see this, consider the first k columns of the matrix:

$$\begin{pmatrix} a_{11} & a_{12} & \cdots & a_{1,k-1} & a_{1k} \\ & a_{22} & \cdots & a_{2,k-1} & a_{2k} \\ & & \ddots & \vdots & \vdots \\ & & & a_{k-1,k-1} & a_{k-1,k} \\ & & & 0 & 0 \\ & & & \vdots & \vdots \\ & & & 0 & 0 \end{pmatrix}.$$

To determine column k as a linear combination of the other columns is equivalent to solving a triangular system of equations

$$\begin{pmatrix} a_{11} & a_{12} & \cdots & a_{1,k-1} \\ & a_{22} & \cdots & a_{2,k-1} \\ & & \ddots & \vdots \\ & & & a_{k-1,k-1} \end{pmatrix} \begin{pmatrix} c_1 \\ c_2 \\ \vdots \\ c_{k-1} \end{pmatrix} = \begin{pmatrix} a_{1k} \\ a_{2k} \\ \vdots \\ a_{k-1,k} \end{pmatrix},$$

which can be done, since the diagonal elements a_{ii}, $i = 1, 2, \ldots, k - 1$, are nonzero.

Thus, we see that, under the assumption that the columns of A are linearly independent (A is nonsingular), Gaussian elimination with partial pivoting can always be carried through (provided that the computations are performed in exact arithmetic). (This argument is not yet complete, however; we must show that the number of linearly independent columns in A does not change during the elimination. This we do in Section 8.6.)

8.5   Permutations, Gauss Transformations and Partitioned Matrices

In the next section, we shall show that Gaussian elimination is equivalent to a factorization of the matrix. For that discussion, we introduce some elementary transformation matrices.

Row pivoting is equivalent to multiplication from the left by a permutation matrix. The simplest permutation matrix, P_{ij}, is constructed by interchanging rows i and j in the identity matrix. Let $n = 5$, for instance. The permutation matrix P_{24} is

$$P_{24} = \begin{pmatrix} 1 & 0 & 0 & 0 & 0 \\ 0 & 0 & 0 & 1 & 0 \\ 0 & 0 & 1 & 0 & 0 \\ 0 & 1 & 0 & 0 & 0 \\ 0 & 0 & 0 & 0 & 1 \end{pmatrix}.$$

We call this a **simple permutation matrix**. Another name is **transposition matrix**.

When multiplying a vector x by P_{ij}, the components x_i and x_j are interchanged:

$$P_{ij}x = \begin{pmatrix} \vdots \\ x_{i-1} \\ x_j \\ x_{i+1} \\ \vdots \\ x_{j-1} \\ x_i \\ x_{j+1} \\ \vdots \end{pmatrix}.$$

Similarly, when we multiply a matrix A by P_{ij} from the left, rows i and j

are interchanged:

$$P_{ij}A = \begin{pmatrix} & & \vdots & & \\ a_{j1} & a_{j2} & \cdots & a_{jn} \\ \vdots & \vdots & & \vdots \\ a_{i1} & a_{i2} & \cdots & a_{in} \\ & & \vdots & & \end{pmatrix}.$$

We immediately see that

$$P_{ij}^{-1} = P_{ij}.$$

Products of simple permutation matrices

$$P_{i_1 j_1} P_{i_2 j_2} \cdots P_{i_k j_k}$$

are also called permutation matrices. (A permutation matrix can be defined as a matrix P such that Px is a reordering of the components of the vector x. It can be shown that every permutation matrix is a product of simple permutation matrices.)

While multiplication of a matrix A from the left by a permutation matrix P is equivalent to a permutation of the rows of A, the corresponding permutation of the columns of A is obtained by a multiplication by P^T from the *right*: AP^T.

Apart from permutation matrices, we need **Gauss transformations**. A Gauss transformation L_j is a lower triangular matrix, deviating from the identity matrix only in column j

$$L_j = \begin{pmatrix} 1 & & & & & & & \\ & 1 & & & & & & \\ & & \ddots & & & & & \\ & & & 1 & & & & \\ & & & m_{j+1,j} & 1 & & & \\ & & & m_{j+2,j} & & 1 & & \\ & & & \vdots & & & \ddots & \\ & & & m_{nj} & & & & 1 \end{pmatrix}.$$

When a vector x is multiplied by L_j, we get

$$L_j x = \begin{pmatrix} x_1 \\ \vdots \\ x_j \\ x_{j+1} + m_{j+1,j}\, x_j \\ x_{j+2} + m_{j+2,j}\, x_j \\ \vdots \\ x_n + m_{nj}\, x_j \end{pmatrix}.$$

Thus, we add $m_{\nu j}x_j$ to the components x_ν, $\nu = j+1, j+2, \ldots, n$.

From this, we see that

$$
L_j^{-1} = \begin{pmatrix}
1 & & & & & & \\
& 1 & & & & & \\
& & \ddots & & & & \\
& & & 1 & & & \\
& & & -m_{j+1,j} & 1 & & \\
& & & -m_{j+2,j} & & 1 & \\
& & & \vdots & & & \ddots \\
& & & -m_{nj} & & & & 1
\end{pmatrix},
$$

since if we multiply the vector $y = L_j x$ by this matrix, we now *subtract* $m_{\nu j}x_j$ from the components y_ν, $\nu = j+1, j+2, \ldots, n$.

Gauss transformations have other important properties that we illustrate in a couple of examples.

Example 8.5.1   Let $k > j$. Then

$$
L_j L_k = \begin{pmatrix}
1 & & & & & & & \\
& \ddots & & & & & & \\
& & 1 & & & & & \\
& & m_{j+1,j} & 1 & & & & \\
& & & & \ddots & & & \\
& & & & & 1 & & \\
& & \vdots & & & m_{k+1,k} & 1 & \\
& & \vdots & & & \vdots & & \ddots \\
& & m_{nj} & & & m_{nk} & & 1
\end{pmatrix}.
$$

(Check by performing the multiplication.)

Example 8.5.2   Given a vector

$$
x = \begin{pmatrix} x_1 \\ x_2 \\ \vdots \\ x_n \end{pmatrix},
$$

where $x_1 \neq 0$. Put

$$
L_1 = \begin{pmatrix}
1 & & & & \\
m_{21} & 1 & & & \\
m_{31} & 0 & 1 & & \\
\vdots & \vdots & & \ddots & \\
m_{n1} & 0 & & & 1
\end{pmatrix},
$$

where

$$
m_{i1} = x_i/x_1, \qquad i = 2, 3, \ldots, n.
$$

Then

$$
L_1^{-1} x = \begin{pmatrix}
x_1 \\
x_2 - m_{21}x_1 \\
x_3 - m_{31}x_1 \\
\vdots \\
x_n - m_{n1}x_1
\end{pmatrix} = \begin{pmatrix}
x_1 \\
0 \\
0 \\
\vdots \\
0
\end{pmatrix}.
$$

Similarly, if $x_j \neq 0$ and

$$
L_j = \begin{pmatrix}
1 & & & & & \\
& \ddots & & & & \\
& & 1 & & & \\
& & m_{j+1,j} & 1 & & \\
& & \vdots & & \ddots & \\
& & m_{nj} & & & 1
\end{pmatrix},
$$

where

$$
m_{ij} = x_i/x_j, \qquad i = j+1, j+2, \ldots, n,
$$

then we have

$$
L_j^{-1} x = \begin{pmatrix}
x_1 \\
x_2 \\
\vdots \\
x_j \\
0 \\
\vdots \\
0
\end{pmatrix}.
$$

In the next example, it turns out to be practical to **partition** the matrix into blocks. We illustrate this with 3×3 matrices. Let

$$
A = \begin{pmatrix}
7 & 5 & 0 \\
3 & 6 & 1 \\
3 & 1 & 6
\end{pmatrix}.
$$

Partition A into blocks

$$A = \begin{pmatrix} A_{11} & A_{12} \\ A_{21} & A_{22} \end{pmatrix},$$

where

$$A_{11} = \begin{pmatrix} 7 & 5 \\ 3 & 6 \end{pmatrix}, \quad A_{12} = \begin{pmatrix} 0 \\ 1 \end{pmatrix},$$

$$A_{21} = \begin{pmatrix} 3 & 1 \end{pmatrix}, \quad A_{22} = \begin{pmatrix} 6 \end{pmatrix}.$$

Partition the matrix B so that B_{ij} has the same dimension as A_{ij}:

$$B = \begin{pmatrix} B_{11} & B_{12} \\ B_{21} & B_{22} \end{pmatrix} = \begin{pmatrix} 9 & 5 & 4 \\ 0 & 2 & 1 \\ 6 & 5 & 0 \end{pmatrix}.$$

Matrices can be partitioned in many different ways; often, the diagonal blocks (A_{11}, A_{22}) are taken to be square. In many cases, the multiplication of two partitioned matrices can be interpreted as if the blocks were scalars (this requires the blocks to be of appropriate dimensions, so that all matrix multiplications are defined). In our example, we get

$$
\begin{aligned}
AB &= \begin{pmatrix} A_{11} & A_{12} \\ A_{21} & A_{22} \end{pmatrix} \begin{pmatrix} B_{11} & B_{12} \\ B_{21} & B_{22} \end{pmatrix} \\
&= \begin{pmatrix} A_{11}B_{11} + A_{12}B_{21} & A_{11}B_{12} + A_{12}B_{22} \\ A_{21}B_{11} + A_{22}B_{21} & A_{21}B_{12} + A_{22}B_{22} \end{pmatrix} = \begin{pmatrix} 63 & 45 & 33 \\ 33 & 32 & 18 \\ 63 & 47 & 13 \end{pmatrix}.
\end{aligned}
$$

(Check this by performing the multiplication both ways.)

More generally, it can be shown that if two square matrices are partitioned conformably (i.e., so that all matrix multiplications below are defined):

$$A = \begin{pmatrix} A_{11} & A_{12} & \cdots & A_{1N} \\ A_{21} & A_{22} & \cdots & A_{2N} \\ \vdots & \vdots & & \vdots \\ A_{N1} & A_{N2} & \cdots & A_{NN} \end{pmatrix}, \quad B = \begin{pmatrix} B_{11} & B_{12} & \cdots & B_{1N} \\ B_{21} & B_{22} & \cdots & B_{2N} \\ \vdots & \vdots & & \vdots \\ B_{N1} & B_{N2} & \cdots & B_{NN} \end{pmatrix},$$

then the product $C = AB$ can be partitioned in the same way, and the blocks in C are given by

$$C_{ij} = \sum_{k=1}^{N} A_{ik} B_{kj}.$$

Example 8.5.3 The matrix L_1 can be partitioned

$$L_1 = \begin{pmatrix} 1 & 0 \\ m & I \end{pmatrix},$$

where

$$m = \begin{pmatrix} m_{21} \\ m_{31} \\ \vdots \\ m_{n1} \end{pmatrix}$$

Let

$$P_n = \begin{pmatrix} 1 & 0 \\ 0 & P_{n-1} \end{pmatrix},$$

where P_{n-1} is a permutation matrix of dimension $n-1$. Then

$$P_n^{-1} = \begin{pmatrix} 1 & 0 \\ 0 & P_{n-1}^{-1} \end{pmatrix},$$

and

$$P_n L_1 P_n^{-1} = \begin{pmatrix} 1 & 0 \\ P_{n-1}m & P_{n-1}P_{n-1}^{-1} \end{pmatrix} = \begin{pmatrix} 1 & 0 \\ P_{n-1}m & I \end{pmatrix}.$$

8.6   *LU* Decomposition

We shall now show that Gaussian elimination with partial pivoting, applied to a nonsingular matrix A, is equivalent to a decomposition (factorization)

$$PA = LU,$$

where P is a permutation matrix, L is a lower triangular matrix with ones on the main diagonal, and U is an upper triangular matrix. We start by considering a numerical example. Then we give a constructive proof of the theorem for 3×3 matrices. Finally, we prove the theorem for $n \times n$ matrices by induction.

Example 8.6.1 In Example 8.4.2, we performed Gaussian elimination with partial pivoting on the matrix

$$A = \begin{pmatrix} 0.6 & 1.52 & 3.5 \\ 2 & 4 & 1 \\ 1 & 2.8 & 1 \end{pmatrix}.$$

The upper triangular matrix that is the result of the elimination is

$$U = \begin{pmatrix} 2 & 4 & 1 \\ 0 & 0.8 & 0.5 \\ 0 & 0 & 3 \end{pmatrix}.$$

We shall now insert the multipliers m_{ik} in a lower triangular matrix. In the first step of the elimination, we interchanged rows 1 and 2, and then had $m_{21} = 0.3$ and $m_{31} = 0.5$. These are now inserted into their positions in L:

$$L := \begin{pmatrix} 1 & & \\ 0.3 & 1 & \\ 0.5 & 0 & 1 \end{pmatrix}.$$

But, in the next step, we interchanged rows 2 and 3. If we had done that in the original matrix, then we would have gotten $m_{21} = 0.5$ and $m_{31} = 0.3$. Therefore, we now interchange these elements in L:

$$L := \begin{pmatrix} 1 & & \\ 0.5 & 1 & \\ 0.3 & 0 & 1 \end{pmatrix}.$$

In the second step of the elimination, we had $m_{31} = 0.4$, and we put

$$L := \begin{pmatrix} 1 & & \\ 0.5 & 1 & \\ 0.3 & 0.4 & 1 \end{pmatrix}.$$

The multiplication $L \cdot U$ now gives

$$LU = \begin{pmatrix} 1 & & \\ 0.5 & 1 & \\ 0.3 & 0.4 & 1 \end{pmatrix} \begin{pmatrix} 2 & 4 & 1 \\ 0 & 0.8 & 0.5 \\ 0 & 0 & 3 \end{pmatrix} = \begin{pmatrix} 2 & 4 & 1 \\ 1 & 2.8 & 1 \\ 0.6 & 1.52 & 3.5 \end{pmatrix},$$

i.e., $LU = PA$, where the permutation matrix P is

$$P = \begin{pmatrix} 0 & 1 & 0 \\ 0 & 0 & 1 \\ 1 & 0 & 0 \end{pmatrix}.$$

Consider a 3×3 matrix

$$A = \begin{pmatrix} a_{11} & a_{12} & a_{13} \\ a_{21} & a_{22} & a_{23} \\ a_{31} & a_{32} & a_{33} \end{pmatrix}.$$

We assume that it is nonsingular. Then we know that Gaussian elimination with partial pivoting can be carried through. In the first step, we pivot, if necessary, and then we zero the positions (2,1) and (3,1). Let P_1 be the permutation matrix, and put

$$A' := P_1 A = \begin{pmatrix} a'_{11} & a'_{12} & a'_{13} \\ a'_{21} & a'_{22} & a'_{23} \\ a'_{31} & a'_{32} & a'_{33} \end{pmatrix},$$

where prime denotes matrix elements after the pivoting. Now put

$$L_1 = \begin{pmatrix} 1 & & \\ m_{21} & 1 & \\ m_{31} & 0 & 1 \end{pmatrix},$$

where $m_{21} = a'_{21}/a'_{11}$ and $m_{31} = a'_{31}/a'_{11}$. Multiplication by L_1^{-1} from the left is equivalent to subtracting the multiple m_{21} of the first row from the second row, and subtracting the multiple m_{31} of the first row from the third. Then, the result of the first step of Gaussian elimination is

$$A^{(1)} := L_1^{-1} A'$$

$$= \begin{pmatrix} 1 & & \\ -m_{21} & 1 & \\ -m_{31} & 0 & 1 \end{pmatrix} \begin{pmatrix} a'_{11} & a'_{12} & a'_{13} \\ a'_{21} & a'_{22} & a'_{23} \\ a'_{31} & a'_{32} & a'_{33} \end{pmatrix} = \begin{pmatrix} a'_{11} & a'_{12} & a'_{13} \\ 0 & a_{22}^{(1)} & a_{23}^{(1)} \\ 0 & a_{32}^{(1)} & a_{33}^{(1)} \end{pmatrix},$$

where

$$a_{ij}^{(1)} = a'_{ij} - m_{i1} a'_{1j}, \qquad i = 2,3, \quad j = 2,3.$$

In the next step, we pivot (possibly interchange rows 2 and 3), and then zero position (3, 2):

$$A'' := P_2 A^{(1)} = \begin{pmatrix} a'_{11} & a'_{12} & a'_{13} \\ 0 & a''_{22} & a''_{23} \\ 0 & a''_{32} & a''_{33} \end{pmatrix}, \, ?$$

$$U := A^{(2)} := L_2^{-1} A'' = \begin{pmatrix} 1 & & \\ 0 & 1 & \\ 0 & -m_{32} & 1 \end{pmatrix} A'' = \begin{pmatrix} a'_{11} & a'_{12} & a'_{13} \\ 0 & a''_{22} & a''_{23} \\ 0 & 0 & a_{33}^{(2)} \end{pmatrix},$$

where

$$m_{32} = a''_{32}/a''_{22},$$

$$a_{33}^{(2)} = a''_{33} - m_{32} a''_{23}.$$

We now have

$$U = L_2^{-1} P_2 L_1^{-1} P_1 A,$$

or

$$A = P_1 L_1 P_2 L_2 U,$$

since P_1 and P_2 are simple permutation matrices and $P_1^{-1} = P_1$, $P_2^{-1} = P_2$. Further, if we multiply by $P = P_2 P_1$ from the left, we get

$$PA = P_2 L_1 P_2 L_2 U.$$

But, since the permutation P_2 only affects rows 2 and 3, then P_2 has the structure

$$\begin{pmatrix} 1 & 0 \\ 0 & Q \end{pmatrix},$$

where Q is a permutation matrix of dimension 2. Therefore, we see that (cf. Example 8.5.3)

$$\overline{L_1} = P_2 L_1 P_2 = \begin{pmatrix} 1 & & \\ m'_{21} & 1 & \\ m'_{31} & 0 & 1 \end{pmatrix},$$

where prime denotes that the elements of column 1 may have changed places. Thus, we have

$$PA = \overline{L_1} L_2 U = LU = \begin{pmatrix} 1 & & \\ m'_{21} & 1 & \\ m'_{31} & m_{32} & 1 \end{pmatrix} U$$

(cf. Example 8.5.1).

We have now shown that, in the special case $n = 3$, Gaussian elimination with partial pivoting is equivalent to *LU* decomposition of the original matrix, whose rows may have been permuted. It is now quite obvious that the proof can be generalized.

Theorem 8.6.2   *LU decomposition.* Any nonsingular $n \times n$ matrix A can be decomposed into

$$PA = LU,$$

where P is a permutation matrix, L is a lower triangular matrix with ones on the main diagonal, and U is an upper triangular matrix.

Proof. The theorem is proved by induction. We have already seen that the theorem is true for $n = 3$. Assume that it holds for $n = N-1$. Consider an $N \times N$ matrix A. In the first step of Gaussian elimination with partial pivoting, we multiply

$$A^{(1)} := L_1^{-1} P_1 A,$$

where P_1 is a simple permutation matrix, and L_1 is a Gauss transformation

$$L_1 = \begin{pmatrix} 1 & 0 \\ m_1 & I \end{pmatrix}, \qquad m_1 = \begin{pmatrix} m_{21} \\ m_{31} \\ \vdots \\ m_{N1} \end{pmatrix}.$$

The result is

$$A^{(1)} = \begin{pmatrix} a'_{11} & a'_{12} & \cdots & a'_{1N} \\ 0 & a^{(1)}_{22} & \cdots & a^{(1)}_{2N} \\ \vdots & & & \\ 0 & a^{(1)}_{N2} & \cdots & a^{(1)}_{NN} \end{pmatrix}.$$

Now define the $(N-1) \times (N-1)$ matrix

$$B = \begin{pmatrix} a^{(1)}_{22} & \cdots & a^{(1)}_{2N} \\ \vdots & & \\ a^{(1)}_{N2} & \cdots & a^{(1)}_{NN} \end{pmatrix}.$$

From the induction assumption we have that B can be decomposed into

$$P_B B = L_B U_B,$$

and we then see that

$$A^{(1)} = \begin{pmatrix} 1 & 0 \\ 0 & P_B^{-1} \end{pmatrix} \begin{pmatrix} 1 & 0 \\ 0 & L_B \end{pmatrix} \begin{pmatrix} a'_{11} & a_2^T \\ 0 & U_B \end{pmatrix},$$

(check this by multiplying together), where

$$a_2^T = (a'_{12} \ a'_{13} \ \cdots \ a'_{1N}).$$

Put

$$R = \begin{pmatrix} a'_{11} & a_2^T \\ 0 & U_B \end{pmatrix}.$$

We have now decomposed the original matrix into

$$P_1 A = L_1 \begin{pmatrix} 1 & 0 \\ 0 & P_B^{-1} \end{pmatrix} \begin{pmatrix} 1 & 0 \\ 0 & L_B \end{pmatrix} U.$$

If we now multiply from the left by

$$\begin{pmatrix} 1 & 0 \\ 0 & P_B \end{pmatrix},$$

and put

$$P = \begin{pmatrix} 1 & 0 \\ 0 & P_B \end{pmatrix} P_1,$$

then we get

$$PA = \begin{pmatrix} 1 & 0 \\ 0 & P_B \end{pmatrix} L_1 \begin{pmatrix} 1 & 0 \\ 0 & P_B^{-1} \end{pmatrix} \begin{pmatrix} 1 & 0 \\ 0 & L_B \end{pmatrix} U. \quad \text{ok} \quad (1)$$

According to Example 8.5.3,

$$\overline{L_1} := \begin{pmatrix} 1 & 0 \\ 0 & P_B \end{pmatrix} L_1 \begin{pmatrix} 1 & 0 \\ 0 & P_B^{-1} \end{pmatrix}$$

$$= \begin{pmatrix} 1 & 0 \\ 0 & P_B \end{pmatrix} \begin{pmatrix} 1 & 0 \\ m_1 & I \end{pmatrix} \begin{pmatrix} 1 & 0 \\ 0 & P_B^{-1} \end{pmatrix} = \begin{pmatrix} 1 & 0 \\ P_B m_1 & I \end{pmatrix}, \quad \text{ok} \quad (2)$$

i.e., $\overline{L_1}$ is a Gauss transformation with the same structure as L_1. From Example 8.5.1, we see that

$$\overline{L_1} \begin{pmatrix} 1 & 0 \\ 0 & L_B \end{pmatrix} = \begin{pmatrix} 1 & 0 \\ P_B m_1 & L_B \end{pmatrix}. \quad \text{ok} \quad (3)$$

Now, if we put

$$L = \begin{pmatrix} 1 & 0 \\ P_B m_1 & L_B \end{pmatrix},$$

then we get

$$PA = LU, \quad \text{because } (1) \ b (3)$$

which proves the theorem. ∎

The *LU* decomposition of a matrix is unique in the following sense.

Theorem 8.6.3 If the pivot sequence is given (i.e., the permutation matrix P is given), then the factors L and U are uniquely determined.

Proof. This follows directly from the fact that LU decomposition is equivalent to Gaussian elimination without pivoting applied to the matrix PA. ∎

When the LU decomposition of a matrix A has been computed, then linear systems of equations $Ax = b$ can be solved very easily (and quickly). We have

$$PAx = LUx = Pb,$$

and, if we put $y = Ux$, then the system can be solved in two steps

1. solve $Ly = Pb$;
2. solve Ux = y.

Since the matrices L and U are triangular, this requires approximately n^2 operations (for large n). There are applications where one must solve several systems with the same matrix A and different right hand sides

$$Ax^{(k)} = b^{(k)}, \qquad k = 1, 2, \dots, p,$$

where the right hand sides are not known simultaneously. The most time-consuming part of the work, the LU decomposition (approximately $n^3/3$ operations) is carried out once and for all. Then, for each right hand side, the solution can be computed in approximately n^2 operations.

An example of such an application is structural analysis, where the right hand side is the load on the construction under study, and the solution corresponds to displacements (cf. Chapter 1). In many cases, it is useful to see the result of one computation with a certain load before another load is tried (note, however, that the operation counts above are not generally valid for this application, since in structural analysis the matrix A is, in most cases, a band matrix).

Later in this chapter, we shall give another example of a situation where right hand sides are not known simultaneously.

Note that the matrices L and U can be stored in the memory locations used for the matrix A (since we know that the elements in the main diagonal of L are equal to one, it not necessary to store them). The only extra memory needed is a vector of integers with n components for storing the sequence of pivot rows.

We now modify the code Gaussian Elimination (Section 8.4) so that L and the pivoting information are stored, the latter in the vector of integers p, which is assumed to be initialized to $p_i = i$, $i = 1, 2, \dots, n$.

> **for** $k := 1$ **to** $n - 1$ **do**
> $\quad \nu := \text{indmax}(a, k, n);$
> $\quad p_\nu := p_k;\ p_k := \nu;$
> $\quad \text{swap}(a, k, \nu, n);$
> $\quad$ **for** $i := k + 1$ **to** n **do** $a_{ik} := a_{ik}/a_{kk};$
> $\quad$ **for all** (i, j), $k + 1 \leq i, j \leq n$ **do**
> $\quad\quad a_{ij} := a_{ij} - a_{ik} * a_{kj};$

Note that we no longer operate on the right hand side. Thus, we factor the matrix separately, and solve the system $Ax = b$ using the LU decomposition. When solving the triangular system $Ly = Pb$, we need the vector p, which contains information about the pivot sequence:

> **for** $i := 1$ **to** n **do**
> $\quad s := b_{p_i};$
> $\quad$ **for** $j := 1$ **to** $i - 1$ **do**
> $\quad\quad s := s - a_{ij} * y_j;$
> $\quad y_i := s;$

8.7 Symmetric, Positive Definite Matrices

We have stated earlier that systems with a symmetric, positive definite matrix can be solved without pivoting and without unnecessary loss of accuracy. In this section, we will first show that the LU decomposition can always be computed without pivoting, i.e., that the "natural" pivot elements (a_{kk}) are always nonzero. Then, we show that growth of the magnitude of the matrix elements (in a certain sense) cannot occur. Further, we will show that it is possible to take advantage of symmetry so that the decomposition becomes symmetric, too, and requires half as much work as in the general case.

Theorem 8.7.1 The LU decomposition

$$A = LU,$$

of any symmetric, positive definite matrix A can be computed without pivoting.

Proof. The proof is by induction. The theorem is trivially true when the dimension n is equal to 1. Assume that it is true for $n = N - 1$. Let A be an $N \times N$ matrix. We first show that $a_{11} > 0$. The positive definiteness of A means that $x^T A x > 0$ for all $x \neq 0$. Choose $x = e_1 = (1, 0, \ldots, 0)^T$. Then

$$e_1^T A e_1 = a_{11} > 0,$$

and thus we can use a_{11} as pivot element.

Before continuing the proof, we note that in an analogous manner it can be shown that all the diagonal elements of A are strictly positive.

The first step of Gaussian elimination is equivalent to a decomposition

$$A = L_1 A^{(1)} = \begin{pmatrix} 1 & & & \\ m_{21} & 1 & & \\ \vdots & & \ddots & \\ m_{N1} & & & 1 \end{pmatrix} \begin{pmatrix} a_{11} & a_{12} & \cdots & a_{1N} \\ 0 & a_{22}^{(1)} & \cdots & a_{2N}^{(1)} \\ \vdots & \vdots & & \vdots \\ 0 & a_{N2}^{(1)} & \cdots & a_{NN}^{(1)} \end{pmatrix},$$

where

$$m_{i1} = a_{i1}/a_{11}, \qquad i = 2, 3, \ldots, N,$$
$$a_{ij}^{(1)} = a_{ij} - m_{i1} a_{1j}, \qquad i = 2, 3, \ldots, N, \quad j = 2, 3, \ldots, N.$$

Further, we see that

$$A = L_1 \hat{A} L_1^T,$$

where

$$\hat{A} = \begin{pmatrix} a_{11} & 0 & \cdots & 0 \\ 0 & a_{22}^{(1)} & \cdots & a_{2N}^{(1)} \\ \vdots & \vdots & & \vdots \\ 0 & a_{N2}^{(1)} & \cdots & a_{NN}^{(1)} \end{pmatrix}$$

(check this by performing the multiplication $\hat{A} L_1^T$).

We now show that the matrix

$$B = \begin{pmatrix} a_{22}^{(1)} & \cdots & a_{2N}^{(1)} \\ \vdots & & \vdots \\ a_{N2}^{(1)} & \cdots & a_{NN}^{(1)} \end{pmatrix},$$

is symmetric and positive definite. Since A is symmetric, we have

$$a_{ij}^{(1)} = a_{ij} - \frac{a_{i1}}{a_{11}} \cdot a_{1j} = a_{ji} - \frac{a_{1i}}{a_{11}} \cdot a_{j1} = a_{ji} - m_{j1} \cdot a_{1i} = a_{ji}^{(1)}.$$

Thus, B is symmetric. Now let $z = (z_2, z_3, \ldots, z_N)^T$ be an arbitrary vector, and put

$$\bar{z} = \begin{pmatrix} 0 \\ z \end{pmatrix}.$$

Then

$$z^T B z = \bar{z}^T \hat{A} \bar{z} = \bar{z}^T L_1^{-1} A L_1^{-T} \bar{z} = y^T A y > 0,$$

since A is positive definite and $y = L_1^{-T} \bar{z} \neq 0$. Therefore, B is also positive definite. Note, in particular, that this implies $a_{22}^{(1)} > 0$.

From the induction assumption, B has an LU decomposition

$$B = L_B U_B,$$

and in the same way as in the proof of Theorem 8.6.2, we have now shown that A has the decomposition

$$A = LU,$$

where

$$L = L_1 \begin{pmatrix} 1 & 0 \\ 0 & L_B \end{pmatrix},$$

$$U = \begin{pmatrix} a_{11} & a_{12} & \cdots & a_{1N} \\ 0 & & & \\ \vdots & & U_B & \\ 0 & & & \end{pmatrix},$$

and the theorem is proved.   ∎

In Section 8.4, we saw that loss of accuracy can occur in Gaussian elimination as some matrix elements may become very large in the elimination process. The next theorem shows that such growth cannot happen in the Gaussian elimination of symmetric, positive definite matrices.

Theorem 8.7.2 Let A be a symmetric, positive definite matrix. The largest element in magnitude in A is positive, and it is on the main diagonal. All the matrix elements that occur during the LU decomposition (without pivoting) of A are smaller in magnitude, or equal to the largest element in A.

We defer the proof of this theorem to the end of this section. Instead, we will now show that it is possible to obtain a symmetric decomposition of a symmetric, positive definite matrix. Consider the LU decomposition

$$A = LU,$$

and let

$$U = \begin{pmatrix} u_{11} & u_{12} & \cdots & u_{1n} \\ & u_{22} & \cdots & u_{2n} \\ & & \ddots & \vdots \\ & & & u_{nn} \end{pmatrix}.$$

Now define a diagonal matrix

$$D = \begin{pmatrix} u_{11} & & & \\ & u_{22} & & \\ & & \ddots & \\ & & & u_{nn} \end{pmatrix}.$$

The multiplication $D^{-1}U$ is equivalent to dividing the elements of each row in U by the corresponding diagonal element:

$$D^{-1}U = \begin{pmatrix} u_{11}^{-1} & & & \\ & u_{22}^{-1} & & \\ & & \ddots & \\ & & & u_{nn}^{-1} \end{pmatrix} \begin{pmatrix} u_{11} & u_{12} & \cdots & u_{1n} \\ & u_{22} & \cdots & u_{2n} \\ & & \ddots & \vdots \\ & & & u_{nn} \end{pmatrix}$$

$$= U' = \begin{pmatrix} 1 & u_{12}' & \cdots & u_{1n}' \\ & 1 & \cdots & u_{2n}' \\ & & \ddots & \vdots \\ & & & u_{nn}' \end{pmatrix},$$

where the elements of the product are given by

$$u_{ij}' = u_{ij}/u_{ii}, \qquad i = 1, 2, \ldots, n, \quad j = i, i+1, \ldots, n.$$

We can now write

$$A = LU = LDD^{-1}U = LDU',$$

and, if we use the symmetry of A,

$$LU = A = A^T = (LDU')^T = (U')^T DL^T.$$

But $(U')^T$ is lower triangular with ones on the main diagonal, and DL^T is upper triangular. Since the LU decomposition of a matrix is unique, we must now have

$$(U')^T = L.$$

Thus, the decomposition can be written

$$A = LDR' = LDL^T.$$

This is called the LDL^T decomposition. We have shown

Theorem 8.7.3 LDL^T **decomposition**. Any symmetric, positive definite matrix A has a decomposition

$$A = LDL^T,$$

where L is lower triangular with ones on the main diagonal, and D is a diagonal matrix with positive diagonal elements.

The positiveness of the diagonal elements of D follows from the proof of Theorem 8.7.1.

Example 8.7.4 The positive definite matrix

$$A = \begin{pmatrix} 8 & 4 & 2 \\ 4 & 6 & 0 \\ 2 & 0 & 3 \end{pmatrix},$$

has the LU decomposition

$$A = LU = \begin{pmatrix} 1 & 0 & 0 \\ 0.5 & 1 & 0 \\ 0.25 & -0.25 & 1 \end{pmatrix} \begin{pmatrix} 8 & 4 & 2 \\ 0 & 4 & -1 \\ 0 & 0 & 2.25 \end{pmatrix},$$

and the LDL^T decomposition

$$A = LDL^T, \quad D = \begin{pmatrix} 8 & 0 & 0 \\ 0 & 4 & 0 \\ 0 & 0 & 2.25 \end{pmatrix}.$$

Since A is symmetric, it is only necessary to store the main diagonal and the elements above it, $n(n+1)/2$ matrix elements in all. Exactly the same amount of storage is needed for the LDL^T decomposition. It is also seen that since only half as many elements as in the ordinary LU decomposition need be computed, the amount of work is also halved, approximately $n^3/6$ operations. When the LDL^T decomposition is computed, it is not necessary to first compute the LU decomposition, but the elements in L and D can be computed directly. Consider the identity

$$A = LDL^T,$$

or

$$
A = \begin{pmatrix}
1 & & & & & \\
l_{21} & 1 & & & 0 & \\
\vdots & & \ddots & & & \\
l_{i1} & \cdots & l_{ii-1} & 1 & & \\
& & & & \ddots & \\
l_{n1} & \cdots & & & & 1
\end{pmatrix}
\begin{pmatrix}
d_1 & d_1 l_{21} & \cdots & d_1 l_{j1} & \cdots & d_1 l_{n1} \\
& d_2 & & d_2 l_{j2} & \cdots & d_2 l_{n2} \\
& & \ddots & \vdots & & \\
& & & d_j & & \\
& 0 & & & \ddots & \\
& & & & & d_n
\end{pmatrix}.
$$

We now get (assume that $i > j$)

$$a_{ij} = \sum_{k=1}^{j-1} l_{ik} d_k l_{jk} + d_j l_{ij},$$

and, for $i = j$,

$$a_{jj} = \sum_{k=1}^{j-1} l_{jk}^2 d_k + d_j.$$

Then, assume that we have computed $d_1, d_2, \ldots, d_{j-1}$, and the elements in columns 1 to $j-1$ in L. We can compute d_j from the second equation:

$$d_j = a_{jj} - \sum_{k=1}^{j-1} l_{jk}^2 d_k;$$

and, then, the elements in column j from the first equation:

$$l_{ij} = \left(a_{ij} - \sum_{k=1}^{j-1} l_{ik} d_k l_{jk}\right)/d_j, \qquad i = j+1, j+2, \ldots, n.$$

But, here we note that the factors $d_k l_{jk}$ in the sum do not depend on i. Therefore, we can organize the computation as follows:

$$r_k = d_k l_{jk}, \qquad k = 1, 2, \ldots, j-1,$$

$$d_j = a_{jj} - \sum_{k=1}^{j-1} l_{jk} r_k,$$

$$l_{ij} = \left(a_{ij} - \sum_{k=1}^{j-1} l_{ik} r_k\right)/d_j, \qquad i = j+1, j+2, \ldots, n.$$

Thus, we can compute the elements in L columnwise, from the left to the right (and the elements in D at the same time). The number of operations for computing l_{ij} is $j-1$, and to compute column j (including d_j), we therefore need

$$\sum_{i=j}^{n}(j-1) = (j-1)(n-j)$$

operations (the number of operations to compute the auxiliary quantities r_k is $j-1$, and we can disregard them). The total amount of work is

$$\sum_{j=1}^{n-1}(j-1)(n-j) \approx n\sum_{j=1}^{n-1}j - \sum_{j=1}^{n-1}j^2 \approx n\cdot\frac{n^2}{2} - \frac{n^3}{3} = \frac{n^3}{6}$$

(for large n). We have shown that the following rule of thumb is valid.

The number of operations to compute the LDL^T decomposition of a symmetric, positive definite $n \times n$ matrix is approximately $n^3/6$. The storage requirement is approximately $n^2/2$.

We have shown earlier that the diagonal elements in D are positive. Therefore, we can put

$$D^{1/2} = \begin{pmatrix} \sqrt{d_1} & & & \\ & \sqrt{d_2} & & \\ & & \ddots & \\ & & & \sqrt{d_n} \end{pmatrix},$$

and then we get

$$A = LDL^T = (LD^{1/2})(D^{1/2}L^T) = U^T U,$$

where U is an upper triangular matrix. This variant of the LDL^T decomposition is called the **Cholesky decomposition**, and it can also be computed in $n^3/6$ operations.

It appears that the Cholesky decomposition has a disadvantage in that it requires the computation of square roots. But, in most modern computers, square roots are computed almost as fast as divisions, and therefore, in practice, the Cholesky decomposition can be computed as fast as the LDL^T decomposition.

It is easy to see that Cholesky decomposition is equivalent to "completion of squares" in the quadratic form $x^T A x$:

$$x^T A x = x^T U^T U x = y^T y = \sum_{i=1}^{n} y_i^2,$$

where $y = Ux$.

In many applications where symmetric matrices arise, one knows from the problem context that the matrix is positive definite. E.g., for the boundary value problem

$$Ly = -y'' + p(x)y = f(x), \qquad y(a) = \alpha, \quad y(\beta) = \beta,$$

where $p(x)$ is nonnegative, it can be shown that the differential operator L is positive definite, and when the equation is discretized in a suitable way, then the resulting linear system of equations has a symmetric, positive definite matrix (cf. Chapter 10). The normal equations in the least squares method have a positive definite matrix (cf. Section 8.13 and Chapter 9). When the system $Ax = b$ arises in certain structural mechanics computations, then the matrix A is called a **stiffness matrix** (cf. Section 10.10). Here basic physical principles imply that A is positive definite.

Now, if one has a linear system of equations $Ax = b$ with a symmetric matrix A, but does not know in advance if A is positive definite, then how can one best solve that system? It is, of course, possible to use ordinary Gaussian elimination, and to be sure to avoid unnecessary loss of accuracy, one should use partial pivoting. Then, the solution requires $n^3/3$ operations. However, there is also an LDL^T decomposition for indefinite matrices, and the computation of it requires $n^3/6$ operations. Here, D is not a diagonal matrix but *block diagonal*. To avoid loss of accuracy and to maintain symmetry, one must pivot both rows and columns.

We now return to the proof of Theorem 8.7.2. We need a couple of lemmas.

Lemma 8.7.5   A symmetric matrix A is positive definite if and only if the eigenvalues of A are positive.

Proof.   Choose x equal to the eigenvector belonging to the eigenvalue λ, i.e., $Ax = \lambda x$. Then

$$0 < x^T A x = \lambda x^T x,$$

which shows that $\lambda > 0$. The converse is left as an exercise.   ∎

Lemma 8.7.6 Let A be symmetric and positive definite. Then any submatrix

$$\begin{pmatrix} a_{rr} & a_{rs} \\ a_{sr} & a_{ss} \end{pmatrix}$$

is positive definite.

Proof. Let y be a vector defined by

$$y_j = \begin{cases} \sigma, & \text{for } j = r, \\ \tau, & \text{for } j = s, \\ 0, & \text{otherwise.} \end{cases}$$

Then

$$0 < y^T A y = \sigma^2 a_{rr} + 2\sigma\tau a_{rs} + \tau^2 a_{ss} = (\sigma \quad \tau)^T \begin{pmatrix} a_{rr} & a_{rs} \\ a_{rs} & a_{ss} \end{pmatrix} \begin{pmatrix} \sigma \\ \tau \end{pmatrix},$$

for arbitrary nonzero vectors $(\sigma \quad \tau)^T$. ∎

Proof (Theorem 8.7.2). Let A be symmetric and positive definite. From Lemma 8.7.6, we know that any 2×2 matrix

$$\begin{pmatrix} a_{rr} & a_{rs} \\ a_{rs} & a_{ss} \end{pmatrix}$$

is positive definite. Then, Lemma 8.7.5 implies that the eigenvalues of this matrix are positive. The eigenvalues are the roots of the second degree equation

$$\lambda^2 - (a_{rr} + a_{ss})\lambda + (a_{rr}a_{ss} - a_{rs}^2) = 0,$$

and the product of the roots must be positive, hence

$$a_{rr}a_{ss} - a_{rs}^2 > 0.$$

We therefore see that

$$|a_{rs}| < \max(a_{rr}, a_{ss}),$$

and we have shown that the largest element (in magnitude) in A is on the main diagonal, and is positive (see the proof of Theorem 8.7.1).

Next, consider the first step of the factorization of A. As in the proof of Theorem 8.7.1, we have

$$A = L_1 \begin{pmatrix} a_{11} & a_{12} & \cdots & a_{1n} \\ 0 & & & \\ \vdots & & B & \\ 0 & & & \end{pmatrix}.$$

Since B is symmetric and positive definite, its largest element is on the main diagonal. Consider a diagonal element in B:

$$0 < a_{ii}^{(1)} = a_{ii} - m_{i1} \cdot a_{1i}$$

$$= a_{ii} - \frac{a_{i1}a_{1i}}{a_{11}}$$

$$= a_{ii} - \frac{a_{1i}^2}{a_{11}} \le a_{ii}.$$

Since any diagonal element in B is smaller than or equal to the corresponding diagonal element in A, any element in B must be smaller than or equal to the largest element in A. The theorem now follows by induction.   ■

8.8   Band Matrices

In many situations, e.g., boundary value problems for ordinary and partial differential equations, matrices arise where a large proportion of the elements are equal to zero. If the nonzero elements are concentrated around the main diagonal, then the matrix is called a band matrix. More precisely, a matrix A is said to be a **band matrix** if there are natural numbers p and q, such that

$$a_{ij} = 0 \text{ if } j - i > p \text{ or } i - j > q.$$

Example 8.8.1   Let $q = 2, p = 1$. Let A be a band matrix of dimension 6:

$$A = \begin{pmatrix} a_{11} & a_{12} & 0 & 0 & 0 & 0 \\ a_{21} & a_{22} & a_{23} & 0 & 0 & 0 \\ a_{31} & a_{32} & a_{33} & a_{34} & 0 & 0 \\ 0 & a_{42} & a_{43} & a_{44} & a_{45} & 0 \\ 0 & 0 & a_{53} & a_{54} & a_{55} & a_{56} \\ 0 & 0 & 0 & a_{64} & a_{65} & a_{66} \end{pmatrix}.$$

$w = q+p+1$ is called the **band width** of the matrix. From the example, we see that w is the maximal number of nonzero elements in any row of A.

When storing a band matrix, we do not store the elements outside the band. Likewise, when linear systems of equations are solved, one can take advantage of the band structure to reduce the number of operations.

We first consider the case $p = q = 1$. Such a band matrix is called **tridiagonal**. Let

$$A = \begin{pmatrix} \alpha_1 & \beta_1 \\ \gamma_2 & \alpha_2 & \beta_2 \\ & \gamma_3 & \alpha_3 & \beta_3 \\ & & & \ddots & \ddots & \ddots \\ & & & & \gamma_{n-1} & \alpha_{n-1} & \beta_{n-1} \\ & & & & & \gamma_n & \alpha_n \end{pmatrix}.$$

The matrix can be stored in three vectors. In the solution of a tridiagonal system $Ax = b$, it is easy to utilize the structure; we first assume that A is diagonally dominant, so that no pivoting is needed.

LU DECOMPOSITION OF A TRIDIAGONAL MATRIX.

> **for** $k := 1$ **to** $n - 1$ **do**
> $\qquad \gamma_{k+1} := m := \gamma_{k+1}/\alpha_k;$
> $\qquad \alpha_{k+1} := \alpha_{k+1} - m * \beta_k;$

FORWARD SUBSTITUTION FOR THE SOLUTION OF $Ly = b$.

> $y_1 := b_1;$
> **for** $k := 2$ **to** n **do** $y_k := b_k - \gamma_k * y_{k-1};$

BACK SUBSTITUTION FOR THE SOLUTION OF $Ux = y$.

> $x_n := y_n/\alpha_n;$
> **for** $k := n - 1$ **downto** 1 **do** $x_k := (y_k - \beta_k * x_{k+1})/\alpha_k;$

One can save storage space, e.g., by letting y and x overwrite b. As in the case of a full matrix, the factors L and U can be stored in the original storage positions.

The number of operations (multiplications and additions) is approximately $3n$, and the number of divisions is $2n$.

If it is necessary to perform partial pivoting in the solution of a tridiagonal system, then the band structure is destroyed to some extent. Consider a tridiagonal matrix of dimension 5, where $*$ denotes a nonzero element,

$$\begin{pmatrix} * & * \\ * & * & * \\ & * & * & * \\ & & * & * & * \\ & & & * & * \end{pmatrix}.$$

Assume that we must interchange rows 1 and 2. Before the element in position (2,1) is zeroed, we have

$$\begin{pmatrix} * & * & * & & \\ * & * & & & \\ & * & * & * & \\ & & * & * & * \\ & & & * & * \end{pmatrix},$$

and afterwards

$$\begin{pmatrix} * & * & * & & \\ 0 & * & + & & \\ & * & * & * & \\ & & * & * & * \\ & & & * & * \end{pmatrix},$$

where + denotes a new nonzero element. That element arises when we add a multiple of the first row to the second. We therefore see that we get *fill-in* in another diagonal above the main diagonal. In the worst case, the upper triangular matrix that is the result of Gaussian elimination with partial pivoting will have the structure

$$\begin{pmatrix} * & * & * & & \\ & * & * & * & \\ & & * & * & * \\ & & & * & * \\ & & & & * \end{pmatrix}.$$

On the other hand, there is no fill-in in the lower triangular factor.

It can be shown that no fill-in occurs in Gaussian elimination *without pivoting* on a band matrix. Symbolically, we can write:

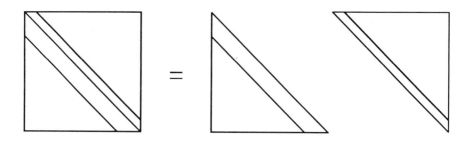

Figure 8.8.2 Gaussian elimination without pivoting.

In Gaussian elimination *with partial pivoting*, the band width of the upper triangular matrix increases. If A has band width $w = q + p + 1$ (q diagonals under the main diagonal and p over), then, with partial pivoting, the factor U will get band width $w_U = p + q + 1$. It is easy to see that no new nonzero elements will be created in L.

While the factors L and U in the LU decomposition of a band matrix A are band matrices, it turns out that as a rule A^{-1} is a dense matrix. Therefore, in most cases the inverse of a band matrix should not be computed explicitly.

Example 8.8.3   Let

$$A = \begin{pmatrix} 4 & 2 & & & \\ 2 & 5 & 2 & & \\ & 2 & 5 & 2 & \\ & & 2 & 5 & 2 \\ & & & 2 & 5 \end{pmatrix}.$$

A has the Cholesky decomposition $A = U^T U$, where

$$U = \begin{pmatrix} 2 & 1 & & & \\ & 2 & 1 & & \\ & & 2 & 1 & \\ & & & 2 & 1 \\ & & & & 2 \end{pmatrix}.$$

The inverse is

$$A^{-1} = \frac{1}{2^{10}} \begin{pmatrix} 341 & -170 & 84 & -40 & 16 \\ -170 & 340 & -168 & 80 & -32 \\ 84 & -168 & 336 & -160 & 64 \\ -40 & 80 & -160 & 320 & -128 \\ 16 & -32 & 64 & -128 & 256 \end{pmatrix}$$

8.9   Computing the Inverse of a Matrix

The inverse of a matrix can be obtained by solving the matrix equation

$$AX = I,$$

where I is the identity matrix. Partition X and I by columns

$$X = (x_{\cdot 1} \ x_{\cdot 2} \ \ldots \ x_{\cdot n}),$$
$$I = (e_1 \ e_2 \ \ldots \ e_n).$$

We then see that the matrix equation can be written

$$AX = (Ax._1 \ Ax._2 \ \ldots \ Ax._n) = (e_1 \ e_2 \ \ldots \ e_n),$$

which means that it is equivalent to n linear systems of equations

$$Ax._k = e_k, \qquad k = 1, 2, \ldots, n.$$

These can be solved easily if we have computed the LU decomposition of A (for simplicity, here we omit permutations):

$$
\begin{array}{ll}
1. & Ly._k = e_k; \\
2. & Ux._k = y._k.
\end{array}
$$

Normally, the solution of a system $Ax._k = e_k$ requires n^2 operations. But, if we consider the first step, we see that we can save some operations by taking advantage of the fact that the first components in e_k are zero:

$$
\begin{pmatrix}
l_{11} & & & & & \\
l_{21} & l_{22} & & & & \\
\vdots & \vdots & \ddots & & & \\
l_{k1} & l_{k2} & \cdots & l_{kk} & & \\
\vdots & \vdots & & \vdots & \ddots & \\
l_{n1} & l_{n2} & \cdots & l_{nk} & \cdots & l_{nn}
\end{pmatrix}
\begin{pmatrix}
y_{1k} \\
y_{2k} \\
\vdots \\
y_{kk} \\
\vdots \\
y_{nk}
\end{pmatrix}
=
\begin{pmatrix}
0 \\
0 \\
\vdots \\
1 \\
\vdots \\
0
\end{pmatrix}.
$$

Since the $k-1$ first components in e_k are zero, the corresponding components of $y._k$ are equal to zero. Solving the system $Ly._k = e_k$ is therefore (in terms of the number of operations) equivalent to solving a triangular system of dimension $n-k+1$, which requires approximately $(n-k+1)^2/2$ operations. No such savings can be made in the solution of $Ux._k = y._k$. The total work to compute the inverse (when the LU decomposition is known already) is then

$$\sum_{k=1}^{n} \left(\frac{(n-k+1)^2}{2} + \frac{n^2}{2} \right) = \frac{1}{2} \sum_{\nu=1}^{n} \nu^2 + n \cdot \frac{n^2}{2} \approx \frac{1}{2} \cdot \frac{n^3}{3} + \frac{n^3}{2} = \frac{2n^3}{3}.$$

We thus have the following rule of thumb.

The computation of the inverse of an $n \times n$ matrix requires approximately n^3 operations.

In practice, one seldom needs the inverse of a matrix. For instance, if we want to compute the matrix product

$$X = A^{-1}B,$$

where B is an $n \times p$ matrix, then this can be done by solving the matrix equation

$$AX = B.$$

But this is equivalent to solving the systems of equations

$$Ax_{.k} = b_{.k}, \qquad k = 1, 2, \ldots, p,$$

where $x_{.k}$ and $b_{.k}$ are column vectors in X and B, respectively. These systems can be solved using the LU decomposition without having to compute the inverse A^{-1}. In certain applications, e.g., in statistics, the elements of the inverse have a special significance, and then sometimes A^{-1} must be computed explicitly.

8.10 Perturbation Theory

In this section, we shall investigate how perturbations in the matrix elements and the right hand side influence the solution of a linear system of equations. Typically such perturbations are measurement errors and rounding errors arising when the matrix and the right hand side are represented in the floating point system.

To be able to estimate errors, we must introduce a measure of the "size" of a vector.

Let $x, y \in R^n$ and $\alpha \in R$. A **vector norm** $\|\cdot\|$ is a mapping $R^n \to R$, with the properties

$$\|x\| \geq 0, \quad \text{for all } x,$$

$$\|x\| = 0 \iff x = 0,$$

$$\|\alpha x\| = |\alpha| \, \|x\|,$$

$$\|x + y\| \leq \|x\| + \|y\|. \qquad \text{(the triangle inequality)}$$

The most common vector norms are

$$\|x\|_1 = \sum_{i=1}^n |x_i|,$$

$$\|x\|_2 = \left(\sum_{i=1}^n x_i^2\right)^{1/2},$$

$$\|x\|_\infty = \max_{1\le i\le n} |x_i|.$$

They are special cases of the ℓ_p norm

$$\|x\|_p = \left(\sum_{i=1}^n |x_i|^p\right)^{1/p}.$$

$\|\cdot\|_2$ is a generalization to R^n of the usual distance in R^3, and is called the **Euclidean norm**. $\|\cdot\|_\infty$ is called the **maximum norm**.

When making statements about vector norms in general (and later matrix norms), we use the notation $\|\cdot\|$.

With norms, we can introduce the concepts of distance and continuity in R^n. Let $\bar{x}$ be an approximation of a nonzero vector x. For a given vector norm $\|\cdot\|$, we define the **absolute error**

$$\|\delta x\| = \|\bar{x} - x\|,$$

and the **relative error**

$$\frac{\|\delta x\|}{\|x\|} = \frac{\|\bar{x} - x\|}{\|x\|}.$$

(Note that δ is *not* a scalar, but δx denotes a vector.)

Example 8.10.1   Let

$$b = \begin{pmatrix} \frac{7}{12} \\ 0.45 \end{pmatrix}, \qquad \bar{b} = \begin{pmatrix} 0.583 \\ 0.45 \end{pmatrix}.$$

Then

$$\delta b = \begin{pmatrix} -0.000333\ldots \\ 0 \end{pmatrix},$$

and

$$\|\delta b\|_\infty = 0.333\ldots \cdot 10^{-3} = \frac{1}{3}\cdot 10^{-3},$$

$$\frac{\|\delta b\|_\infty}{\|b\|_\infty} = \frac{\frac{1}{3}\cdot 10^{-3}}{\frac{7}{12}} = \frac{4}{7}10^{-3}.$$

For an arbitrary vector norm, we can define a corresponding matrix norm.

Let $\|\cdot\|$ be a vector norm. The **corresponding matrix norm** is defined as

$$\|A\| = \sup_{x \neq 0} \frac{\|Ax\|}{\|x\|}.$$

It can (easily) be shown that such a matrix norm satisfies

$$\|A\| \geq 0 \quad \text{for all } A,$$

$$\|A\| = 0 \quad \Leftrightarrow \quad A = 0,$$

$$\|\alpha A\| = |\alpha| \, \|A\|,$$

$$\|A + B\| \leq \|A\| + \|B\|,$$

where α is a scalar. For the matrix norms defined above, the following fundamental inequalities hold.

Lemma 8.10.2 Let $\|\cdot\|$ denote a vector norm and corresponding matrix norm. Then

$$\|Ax\| \leq \|A\| \, \|x\|,$$

$$\|AB\| \leq \|A\| \, \|B\|.$$

Proof. From the definition, we see that

$$\frac{\|Ax\|}{\|x\|} \leq \|A\|,$$

for all $x \neq 0$. The second inequality in the lemma is obtained by using the first twice on $\|ABx\|$. ∎

One can show that

$$\|A\|_2 = (\max_{1 \leq i \leq n} \lambda_i(A^T A))^{1/2},$$

i.e., the square root of the largest eigenvalue of the matrix $A^T A$. Thus, it is rather time-consuming to compute $\|A\|_2$ for a given matrix A. It is much simpler to compute $\|A\|_\infty$.

Lemma 8.10.3

$$\|A\|_\infty = \max_{1\le i\le n}\left(\sum_{j=1}^n |a_{ij}|\right).$$

Proof. Consider the component i in the vector Ax:

$$|(Ax)_i| = \left|\sum_{j=1}^n a_{ij}x_j\right| \le \sum_{j=1}^n |a_{ij}x_j|$$

$$\le \max_{1\le j\le n} |x_j| \sum_{j=1}^n |a_{ij}| = \|x\|_\infty \sum_{j=1}^n |a_{ij}|.$$

Here, we have first used the triangle inequality, and then replaced each component of the vector x by the component largest in magnitude. We now see that

$$\|Ax\|_\infty = \max_{1\le i\le n} |(Ax)_i| \le \|x\|_\infty \max_{1\le i\le n}\left(\sum_{j=1}^n |a_{ij}|\right),$$

so that

$$\|A\|_\infty \le \max_{1\le i\le n}\left(\sum_{j=1}^n |a_{ij}|\right).$$

We shall now prove that there is a vector x such that

$$\frac{\|Ax\|_\infty}{\|x\|_\infty} = \max_{1\le i\le n}\left(\sum_{j=1}^n |a_{ij}|\right).$$

Choose ν such that

$$\sum_{j=1}^n |a_{\nu j}| = \max_{1\le i\le n}\left(\sum_{j=1}^n |a_{ij}|\right),$$

and put

$$x_j = \text{sign}(a_{\nu j}), \qquad j = 1,2,\ldots,n.$$

Then

$$(Ax)_\nu = \sum_{j=1}^{n} a_{\nu j} x_j = \sum_{j=1}^{n} |a_{\nu j}|,$$

and, since $\|x\|_\infty = 1$, the lemma follows. ∎

Example 8.10.4 Let

$$A = \begin{pmatrix} 2 & -18 \\ 6 & 8 \end{pmatrix}.$$

Then $\|A\|_\infty = \max\{2+18, 6+8\} = 20$.

The maximum norm of a matrix is easy to compute, and we are going to use it in the numerical examples in the sequel. The general results that we will show, are valid also for the other norms that have been defined above.

Now, consider a linear system of equations

$$Ax = b.$$

We refer to this as the **exact** system with the **exact** solution x. If we have a perturbation δb of the right hand side, we will also get a perturbed solution

$$A(x + \delta x) = b + \delta b.$$

We will now estimate the relative error $\|\delta x\| / \|x\|$. Since $Ax = b$, we have

$$A\,\delta x = \delta b,$$

or

$$\delta x = A^{-1}\delta b.$$

We take norms and use Lemma 8.10.2:

$$\|\delta x\| = \left\|A^{-1}\delta b\right\| \le \left\|A^{-1}\right\| \|\delta b\|,$$

which gives an estimate of the absolute error. Similarly, we get

$$\|b\| = \|Ax\| \le \|A\| \|x\|,$$

which can be written

$$\frac{1}{\|x\|} \le \|A\| \frac{1}{\|b\|}.$$

Hence,

$$\frac{\|\delta x\|}{\|x\|} \leq \frac{\left\|A^{-1}\right\| \|\delta b\|}{\|x\|} \leq \|A\| \left\|A^{-1}\right\| \frac{\|\delta b\|}{\|b\|}.$$

The relative perturbation of b is magnified by a factor $\|A\| \left\|A^{-1}\right\|$. This is called the condition number of the matrix.

The condition number of a nonsingular matrix A is defined as

$$\kappa(A) = \|A\| \left\|A^{-1}\right\|.$$

The condition number depends on which matrix norm has been chosen (but only to a minor extent: the different condition numbers of a matrix are all of the same order of magnitude). Sometimes, one wants to emphasize that a certain matrix norm has been used, and then one can, e.g., write

$$\kappa_\infty(A) = \|A\|_\infty \left\|A^{-1}\right\|_\infty.$$

Example 8.10.5 Let

$$b = \begin{pmatrix} \frac{7}{12} \\ 0.45 \end{pmatrix}, \qquad \bar{b} = \begin{pmatrix} 0.583 \\ 0.45 \end{pmatrix}.$$

Then we have

$$\frac{\|\delta b\|_\infty}{\|b\|_\infty} = \frac{\frac{1}{3} \cdot 10^{-3}}{\frac{7}{12}} = \frac{4}{7} \cdot 10^{-3}$$

(see Example 8.10.1). Let

$$A = \begin{pmatrix} \frac{1}{3} & \frac{1}{4} \\ \frac{1}{4} & \frac{1}{5} \end{pmatrix}, \qquad A^{-1} = \begin{pmatrix} 48 & -60 \\ -60 & 80 \end{pmatrix}.$$

The condition number of A is

$$\kappa_\infty(A) = \|A\|_\infty \left\|A^{-1}\right\|_\infty = \frac{7}{12} \cdot 140 \approx 81.7.$$

The system $Ax = b$ has the exact solution

$$x = \begin{pmatrix} 1 \\ 1 \end{pmatrix}.$$

The solution of the perturbed system $Ax = \bar{b}$ is

$$\bar{x} = \begin{pmatrix} 0.984 \\ 1.020 \end{pmatrix},$$

and

$$\frac{\|\bar{x} - x\|_\infty}{\|x\|_\infty} = 0.02.$$

According to the theory, we get the following bound for the relative perturbation in the solution:

$$\frac{\|\delta x\|_\infty}{\|x\|_\infty} \leq \kappa_\infty(A) \frac{\|\delta b\|_\infty}{\|b\|_\infty} \leq 81.7 \cdot \frac{4}{7} \cdot 10^{-3} \leq 0.047.$$

Thus, the error estimate gives a a somewhat large value, but it is quite realistic.

From the derivation and the example above, we see that the condition number is a measure of the sensitivity of the solution of a linear system to perturbations of the right hand side. The following theorem shows that the condition number can also be used to estimate the sensitivity to perturbations of the matrix.

Theorem 8.10.6   Assume that A is nonsingular and that

$$\|\delta A\| \, \|A^{-1}\| = r < 1.$$

Then the matrix $A + \delta A$ is nonsingular, and

$$\|(A + \delta A)^{-1}\| \leq \frac{\|A^{-1}\|}{1 - r}.$$

The solution of the perturbed system

$$(A + \delta A)y = b + \delta b$$

satisfies

$$\frac{\|y - x\|}{\|x\|} \leq \frac{\kappa(A)}{1 - r} \left(\frac{\|\delta A\|}{\|A\|} + \frac{\|\delta b\|}{\|b\|} \right).$$

Proof.   We do not give all details in the proof. The matrix $I + F$, where $F = (\delta A)A^{-1}$, has the inverse

$$\sum_0^\infty (-1)^k F^k,$$

and

$$\left\|\sum_0^\infty (-1)^k F^k\right\| \leq \sum_0^\infty \|F\|^k = \frac{1}{1-r}.$$

Therefore,

$$\left\|(A + \delta A)^{-1}\right\| = \left\|A^{-1}(I + (\delta A)\,A^{-1})^{-1}\right\| \leq \frac{\|A^{-1}\|}{1-r}.$$

Further, we get

$$y - x = \left((A + \delta A)^{-1} - A^{-1}\right)b + (A + \delta A)^{-1}\delta b.$$

But,

$$\begin{aligned}
(A + \delta A)^{-1} - A^{-1} &= \left((A + \delta A)^{-1}A - I\right)A^{-1} \\
&= \left((A + \delta A)^{-1}(A + \delta A) - (A + \delta A)^{-1}\delta A - I\right)A^{-1} \\
&= -(A + \delta A)^{-1}(\delta A)\,A^{-1},
\end{aligned}$$

which gives

$$\begin{aligned}
y - x &= (A + \delta A)^{-1}\left((-\delta A)A^{-1}b + \delta b\right) \\
&= (A + \delta A)^{-1}\left((-\delta A)x + \delta b\right).
\end{aligned}$$

Taking norms, we get

$$\begin{aligned}
\|y - x\| &\leq \left\|(A + \delta A)^{-1}\right\|\left(\|\delta A\|\,\|x\| + \|\delta b\|\right) \\
&\leq \frac{\|A^{-1}\|}{1-r}\left(\|\delta A\|\,\|x\| + \|\delta b\|\right),
\end{aligned}$$

and, further,

$$\frac{\|y - x\|}{\|x\|} \leq \frac{\|A^{-1}\|\,\|A\|}{1-r}\left(\frac{\|\delta A\|}{\|A\|} + \frac{\|\delta b\|}{\|b\|}\right),$$

where the second term is estimated using the first inequality of Lemma 8.10.2. ∎

It can be shown that, for condition numbers corresponding to ℓ_p norms, we have $\kappa_p(A) \geq 1$ for all (nonsingular) matrices A.

A matrix with a small condition number is called **well-conditioned**. An orthogonal matrix Q (i.e., a matrix satisfying $Q^T Q = I$) is as well-conditioned as any matrix can be, since $\kappa_2(Q) = 1$.

A matrix with a large condition number is said to be **ill-conditioned**. The condition number gives an estimate of the accuracy that can be expected when a linear system $Ax = b$ is solved. If, for instance, $\kappa(A) = 10^3$, and there are relative errors in the matrix and the right hand side of the order of magnitude 10^{-6}, then one can expect the solution to have approximately three significant digits.

Example 8.10.7   The matrix

$$A = \begin{pmatrix} 1 & 1 \\ 1 & 1.0001 \end{pmatrix}$$

has the condition number

$$\kappa_\infty(A) \approx 4 \cdot 10^4.$$

The matrix is rather ill-conditioned. This is equivalent to the column vectors being almost linearly dependent. In this example, it is clearly seen that the column vectors are almost parallel.

It is tempting to believe that the determinant of a matrix is a measure of its conditioning. Since the determinant of a singular matrix is equal to zero, it seems plausible that, if $\det(A)$ is small, then A is ill-conditioned (almost singular). Conversely, if the determinant is large, then it seems that the matrix must be well-conditioned. It turns out that the determinant is completely useless for investigating the conditioning of a matrix, as can be seen from the following example.

Example 8.10.8   Let A_n be a triangular $n \times n$ matrix,

$$A_n = \begin{pmatrix} 1 & -1 & -1 & \cdots & -1 & -1 \\ & 1 & -1 & \cdots & -1 & -1 \\ & & 1 & \cdots & -1 & -1 \\ & & & \ddots & & \\ & & & & 1 & -1 \\ & & & & & 1 \end{pmatrix}.$$

Here, $\kappa_\infty(A_n) = n \cdot 2^{n-1}$, while the determinant is $\det(A_n) = 1$. On the other hand, the diagonal matrix

$$D_n = \begin{pmatrix} 10^{-2} & & & \\ & 10^{-2} & & \\ & & \ddots & \\ & & & 10^{-2} \end{pmatrix},$$

has $\det(D_n) = 10^{-2n}$, but the condition number is $\kappa_\infty(D_n) = 1$.

As a rule, one does not compute the condition number when solving a system $Ax = b$, since this requires the computation of the inverse A^{-1}, which takes about n^3 operations. However, it is possible to *estimate* the condition number (at a cost of $O(n^2)$ operations). We discuss this briefly in Section 8.12.

8.11   Rounding Errors in Gaussian Elimination

From the section on floating point arithmetic in Chapter 2, we know that any real number (representable in the floating point system) is represented with a relative error not exceeding the unit roundoff μ. This fact can also be stated

$$\mathrm{fl}[x] = x(1 + \epsilon), \qquad |\epsilon| \le \mu.$$

When representing the elements of a matrix A and a vector b in the floating point system, there arise errors:

$$\mathrm{fl}[a_{ij}] = a_{ij}(1 + \epsilon_{ij}), \qquad |\epsilon_{ij}| \le \mu,$$

and analogously for b. Therefore, we can write

$$\mathrm{fl}[A] = A + \delta A,$$
$$\mathrm{fl}[b] = b + \delta b,$$

where

$$\|\delta A\|_\infty \le \mu \|A\|_\infty,$$
$$\|\delta b\|_\infty \le \mu \|b\|_\infty.$$

If, for the moment, we assume that no further rounding errors arise during the solution of the system $Ax = b$, we see that the computed solution $\hat{x}$ is the exact solution of

$$(A + \delta A)\hat{x} = b + \delta b.$$

From Theorem 8.10.6, we then get

$$\frac{\|\hat{x} - x\|_\infty}{\|x\|_\infty} \leq \frac{\kappa_\infty(A)}{1 - r} 2\mu$$

(provided that $r = \mu \kappa_\infty(A) < 1$).

This is an example of *backward error analysis* (see Chapter 2).

Backward error analysis. The computed solution $\hat{x}$ is the exact solution of a **perturbed** problem

$$(A + \delta A)\hat{x} = b + \delta b.$$

Using perturbation theory, we can estimate the error in $\hat{x}$.

We can also analyze how rounding errors in Gaussian elimination affect the result. The following theorem holds.

Theorem 8.11.1   Assume that we use a floating point system with unit roundoff μ. Let $\hat{L}$ and $\hat{R}$ be the triangular factors obtained from Gaussian elimination with partial pivoting, applied to the matrix A. Further, assume that $\hat{x}$ is computed using forward and back substitution:

$$\hat{L}\hat{y} = Pb, \quad \hat{R}\hat{x} = \hat{y}.$$

Then $\hat{x}$ is the exact solution of a system

$$(A + \delta A)\hat{x} = b,$$

where

$$\|\delta A\|_\infty \leq k_2(n)\mu \|A\|_\infty, \qquad k_2(n) = (n^3 + 3n^2)g_n,$$

$$g_n = \frac{\max_{i,j,k}|\hat{a}_{ij}^{(k)}|}{\max_{i,j}|a_{ij}|}$$

($\hat{a}_{ij}^{(k)}$ are the elements computed in step $k - 1$ of the elimination procedure).

The proof of the theorem is rather technical, and is omitted.

We observe that g_n depends on the growth of the matrix elements during the Gaussian elimination, and not explicitly on the magnitude of the multipliers. g_n can be computed, and this way an a *posteriori* estimate of the rounding errors can be obtained.

A *priori* (in advance), one can show that $g_n \leq 2^{n-1}$, and matrices can be constructed where in fact the element growth is that serious (note that $g_{31} = 2^{30} \approx 10^9$). Therefore, it may seem necessary always to use full pivoting (the corresponding estimate for full pivoting is $g_n \leq 1.8 n^{0.25 \log n}$; $g_{31} \leq 34.4$). In practice, however, g_n is seldom larger than 8 in partial pivoting.

We want to emphasize that the estimate in the theorem, in almost all cases, is much too pessimistic. In order to have equality, all rounding errors must be maximally large (equal to μ), and their accumulated effect must be maximally unfavourable. Therefore, the main object of this type of a priori error analysis is not to give error estimates for the solution of linear systems, but rather to expose potential instabilities of algorithms, and thus provide a basis for comparing different algorithms.

In Section 8.7, we showed that growth of the matrix elements cannot occur when A is symmetric and positive definite. For such matrices, we thus have $g_n = 1$.

We close this section with a few examples, which further illustrate perturbation and error analysis.

Example 8.11.2   In Section 8.4, we considered a matrix

$$A = \begin{pmatrix} \epsilon & 1 \\ 1 & 1 \end{pmatrix}.$$

This matrix has the inverse

$$A^{-1} = \frac{1}{\epsilon - 1} \begin{pmatrix} 1 & -1 \\ -1 & \epsilon \end{pmatrix},$$

and the condition number

$$\kappa_\infty(A) \approx 4,$$

if ϵ is small. Thus the *problem* of solving $Ax = b$ is well-conditioned.

If we use Gaussian elimination *without* pivoting, we get an enormous growth in the matrix elements:

$$\begin{pmatrix} \epsilon & 1 \\ 1 & 1 \end{pmatrix} = \begin{pmatrix} 1 & 0 \\ 1/\epsilon & 1 \end{pmatrix} \begin{pmatrix} \epsilon & 1 \\ 0 & 1 - 1/\epsilon \end{pmatrix}.$$

According to Theorem 8.11.1, we may now expect large errors in the solution. The factors L and U are very ill-conditioned:

$$\kappa_\infty(L) \approx \frac{2}{\epsilon}, \quad \kappa_\infty(U) \approx \frac{1}{\epsilon}.$$

The *algorithm* is unstable.

The *LU* decomposition of the same matrix with permuted rows is

$$\begin{pmatrix} 1 & 1 \\ \epsilon & 1 \end{pmatrix} = \begin{pmatrix} 1 & 0 \\ \epsilon & 1 \end{pmatrix} \begin{pmatrix} 1 & 1 \\ 0 & 1 - \epsilon \end{pmatrix}.$$

No growth of matrix elements occurs, and the factors are very well conditioned:

$$\kappa_\infty(L) \approx 1, \quad \kappa_\infty(U) \approx 4.$$

Example 8.11.3   The matrix

$$A = \begin{pmatrix} 2 & 1 \\ 1 & 2 \end{pmatrix},$$

is symmetric and positive definite (show this!). It is well-conditioned:

$$\kappa_\infty(A) = 3.$$

The *LDL^T* decomposition is

$$A = \begin{pmatrix} 1 & 0 \\ \frac{1}{2} & 1 \end{pmatrix} \begin{pmatrix} 2 & 0 \\ 0 & \frac{3}{2} \end{pmatrix} \begin{pmatrix} 1 & \frac{1}{2} \\ 0 & 1 \end{pmatrix}.$$

No growth of matrix elements occurs.

Next, consider the matrix

$$B = \begin{pmatrix} \epsilon & 1 \\ 1 & 1/\epsilon + 1 \end{pmatrix}.$$

B is also symmetric and positive definite. It is ill-conditioned:

$$\kappa_\infty(A) \approx \frac{1}{\epsilon^2},$$

and has the *LDL^T* decomposition

$$B = \begin{pmatrix} 1 & 0 \\ 1/\epsilon & 1 \end{pmatrix} \begin{pmatrix} \epsilon & 0 \\ 0 & 1 \end{pmatrix} \begin{pmatrix} 1 & 1/\epsilon \\ 0 & 1 \end{pmatrix}.$$

No growth of matrix elements occurs. The problem of solving $Bx = c$ is indeed ill-conditioned, but the algorithm (Gaussian elimination *without* pivoting) does not introduce any *unnecessary* loss of accuracy.

8.12   Iterative Refinement and Condition Number Estimation

We have seen that the solution of $Ax = b$ obtained from a computer is only an approximation of the exact solution, due to rounding errors introduced firstly at the representation of A and b in a floating point system, and secondly during Gaussian elimination. The best one could ever hope for is to get the solution (rounded to machine precision) of the perturbed system

$$(A + \delta A)x = b + \delta b,$$

where δA and δb correspond to the rounding error in the representation of A and b in the floating point system. We shall now demonstrate a method to solve $Ax = b$, such that this maximal accuracy is obtained.

We assume that an LU decomposition of the matrix A has been computed; due to rounding errors, it is not exact:

$$\hat{L}\hat{U} = PA + E,$$

where $\|E\|$ can be estimated as in Theorem 8.11.1. Let $\hat{x}$ be the approximate solution computed using forward and back substitution:

$$\hat{L}\hat{y} = Pb, \quad \hat{U}\hat{x} = \hat{y}.$$

Now, if we compute the **residual vector**

$$r = b - A\hat{x},$$

and solve

$$Az = r,$$

then $\hat{x} + z$ is the *exact* solution of $Ax = b$, provided that we compute without rounding errors. This cannot be done, however, and we must modify the procedure so that it works also in floating point arithmetic.

Since the rounding errors in the computed solution $\hat{x}$ are likely to be small, we can expect the residual vector r to be small too, probably of the order of magnitude of the unit roundoff. But the *errors* in the residual vector are also of same order of magnitude. Therefore, in order to separate, so to say, the real information from rounding errors, we should compute r in *double precision*. Note that the original matrix A must be used in the computation of r.

In some cases, it may be necessary to repeat the procedure. We get the following algorithm:

ITERATIVE REFINEMENT.

repeat until convergence
$r := b - Ax;$ (double precision)
Solve $\hat{L}y = Pr;$
Solve $\hat{R}z = y;$
$x := x + z;$

If the matrix A is not extremely ill-conditioned, then one step of iterative refinement is enough to get the solution to full machine precision (cf. the examples below). Note that only the computation of the residual vector need be done in double precision. Since we use the LU decomposition, each iteration requires only n^2 operations.

Example 8.12.1 $Ax = b$, where

$$A = \begin{pmatrix} 7 & 6.9 \\ 4 & 4 \end{pmatrix}, \qquad b = \begin{pmatrix} 34.7 \\ 20 \end{pmatrix},$$

is solved in the floating point system $(10, 4, -99, 99)$. A has the inverse

$$A^{-1} = \begin{pmatrix} 10 & -17.25 \\ -10 & 17.5 \end{pmatrix},$$

and the condition number $\kappa_\infty(A) \approx 382$. The LU decomposition is

$$\hat{L} \cdot \hat{R} = \begin{pmatrix} 1 & 0 \\ 5.7143 \cdot 10^{-1} & 1 \end{pmatrix} \begin{pmatrix} 7 & 6.9 \\ 0 & 5.7133 \cdot 10^{-2} \end{pmatrix}$$

(the computations were performed in extended precision, and then l_{21} and r_{22} were rounded). We solve $Ax = b$ by forward and backward substitution:

$$\hat{x} = \begin{pmatrix} 2.0003 \\ 2.9997 \end{pmatrix}.$$

The residual is computed in double precision

$$r = b - A\hat{x} = \begin{pmatrix} 34.7 - 7 \cdot 2.0003 - 6.9 \cdot 2.9997 \\ 20 - 4 \cdot 2.0003 - 4 \cdot 2.9997 \end{pmatrix} = \begin{pmatrix} 0.0000 \\ 1.9003 \cdot 10^{-5} \end{pmatrix},$$

and $Az = r$ is solved using forward and backward substitution:

$$z = \begin{pmatrix} -3.2786 \cdot 10^{-4} \\ 3.3261 \cdot 10^{-4} \end{pmatrix},$$

so that

$$\hat{x} := \hat{x} + z = \begin{pmatrix} 2.0000 \\ 3.0000 \end{pmatrix}.$$

This is the solution to full machine precision.

Example 8.12.2   The system $Ax = b$, where

$$A = \begin{pmatrix} 1.2971 & 0.8648 \\ 0.2161 & 0.1441 \end{pmatrix}, \quad b = \begin{pmatrix} 2.1619 \\ 0.3602 \end{pmatrix},$$

is solved using a computer that implements the IEEE floating point standard, i.e., the floating point system $(2, 23, -126, 127)$.

The matrix has the condition number $\kappa_\infty(A) \approx 1.1 \cdot 10^5$. The approximate solution computed by Gaussian elimination and backward substitution is

$$\hat{x} = \begin{pmatrix} 1.001029 \\ 0.9984569 \end{pmatrix}.$$

After one step of iterative refinement, we have

$$\hat{x} := \hat{x} + z = \begin{pmatrix} 1.001043 \\ 0.9984353 \end{pmatrix},$$

which is also the result after two steps of iterative refinement.

This numerical example illustrates what we discussed at the beginning of Section 8.11.

The system $Ax = b$ has the exact solution $x = (1,1)^T$. When we store A and b, rounding errors are introduced, so that the stored system has the solution (rounded to seven significant digits)

$$\hat{x} = \begin{pmatrix} 1.001043 \\ 0.9984353 \end{pmatrix}.$$

According to the discussion in Section 8.11, the two solutions may differ:

$$\frac{\|\hat{x} - x\|_\infty}{\|x\|_\infty} \leq 2\mu \frac{\kappa_\infty(A)}{1 - r}.$$

Here, we have $\mu = \frac{1}{2} \cdot 2^{-23} \approx 5.96 \cdot 10^{-8}$, $\kappa_\infty(A) \approx 1.1 \cdot 10^5$ and $r = \mu\kappa_\infty(A) \approx 6.56 \cdot 10^{-3}$. The theoretical estimate is

$$\frac{\|\hat{x} - x\|_\infty}{\|x\|_\infty} \leq 1.32 \cdot 10^{-2}.$$

The true difference is

$$\frac{\|\hat{x} - x\|_\infty}{\|x\|_\infty} \approx 1.6 \cdot 10^{-3}.$$

When we solve a system of equations $Ax = b$ and want to find out how accurate the solution is, we need to estimate the condition number $\|A\|_\infty \|A^{-1}\|_\infty$.

The computation of A^{-1} requires significantly more work (n^3 operations) than solving $Ax = b$ ($n^3/3$ operations). Therefore, we want to avoid this. But we can estimate $\|A^{-1}\|_\infty$ rather simply. For an arbitrary right hand side d, the following implication holds:

$$Ay = d \;\Rightarrow\; y = A^{-1}d \;\Rightarrow\; \|y\|_\infty \le \|A^{-1}\|_\infty \|d\|_\infty .$$

Thus, for an arbitrary vector d and $y = A^{-1}d$, we have

$$\|A^{-1}\| \ge \frac{\|y\|_\infty}{\|d\|_\infty}.$$

Now, the idea is to choose d so that $\|d\|_\infty = 1$ (i.e., $d_i = \pm 1$), and so that $\|y\|_\infty$ becomes as large as possible.

We consider the special case $A = U$, where U is an upper triangular matrix. Backward substitution for the solution of $Uy = d$ can be written

$$y_k = \left(d_k - \sum_{i=k+1}^{n} u_{ki}y_i\right)/u_{kk}, \qquad k = n, n-1, \ldots, 1.$$

But, it is also possible to organize the algorithm so that the sums are accumulated in the course of the computation of y_k:

> **for** $i := 1$ **to** n **do** $p_i := 0$;
> **for** $k := n$ **downto** 1 **do**
> $\qquad y_k := (d_k - p_k)/u_{kk}$;
> $\qquad$ **for** $i := 1$ **to** $k-1$ **do**
> $\qquad\qquad p_i := p_i + u_{ik}y_k$;

(Apply this to a small example and see that it is equivalent to the usual algorithm.)

Normally, we use backward substitution for a *given* system $Uy = d$, but here we will *choose* the components of d so that y_k becomes as large as possible. Since we have the constraint $\|d\|_\infty = 1$, we see that we shall put d_k equal to $+1$ or -1 depending on which of $(1-p_k)/u_{kk}$ and $(-1-p_k)/u_{kk}$ is largest. In other words, the assignment statement shall be

$$y_k := \left(-\operatorname{sign}(p_k) - p_k\right)/u_{kk};$$

Example 8.12.3   Let

$$U = \begin{pmatrix} 1 & -1 & -1 & -1 \\ 0 & 1 & -1 & -1 \\ 0 & 0 & 1 & -1 \\ 0 & 0 & 0 & 1 \end{pmatrix}.$$

If we apply the algorithm to this matrix, we get

$$
y = \begin{pmatrix} 8 \\ 4 \\ 2 \\ 1 \end{pmatrix}, \qquad d = \begin{pmatrix} 1 \\ 1 \\ 1 \\ 1 \end{pmatrix},
$$

so that $\|y\|_\infty = 8$. The inverse is

$$
U^{-1} = \begin{pmatrix} 1 & 1 & 2 & 4 \\ 0 & 1 & 1 & 2 \\ 0 & 0 & 1 & 1 \\ 0 & 0 & 0 & 1 \end{pmatrix},
$$

which has $\|R^{-1}\|_\infty = 8$.

In most cases, this algorithm does not give as good an estimate as in this example.

The algorithm can be modified so that the accumulated sums p_k also become large.

To estimate the condition number of a matrix A, whose LU decomposition $PA = LU$ is known, the following method is used.

1. Apply the modified algorithm to obtain a large norm solution of $U^T y = d$.
2. Solve $L^T r = y$,
$$Lw = Pr,$$
$$Uz = w.$$
3. $\hat{\kappa} := \|A\|_\infty \, \|z\|_\infty \, / \, \|r\|_\infty$.

It is beyond the scope of this book to analyze the details of this algorithm. It is sufficient to note that, in practice, a good estimate of the order of magnitude of the condition number is obtained. Since we only solve triangular systems, the total work for computing $\hat{\kappa}$ is $O(n^2)$ operations.

8.13 Overdetermined Linear Systems

In this section, we present the method of least squares in a matrix–vector language. The theory is analogous to that given in Chapter 9, where we use a function theoretic language. There are overlappings between these sections, and they can be read independently. We start with an example.

Example 8.13.1   Assume that we want to determine the spring constant for a spring by hanging a weight to the spring and measuring its length. Hooke's law states that the length of the spring is a linear function of the force:

$$e + \kappa F = l,$$

where e is the length of the spring when no force is applied, κ is a constant, F is the force and l is the length. The spring constant k is defined as $k = \kappa^{-1}$. We have measured the following values:

F	l
1	7.97
2	10.2
3	14.2
4	16.0
5	21.2

According to the theory, the measured values should lie on a straight line, but, from Figure 8.13.2, we see that they do not.

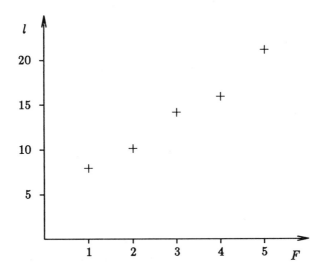

Figure 8.13.2   The length of the spring (l) as a function of the force (F).

The reason for this is not primarily that the relation between the force and the length is more complicated than a linear function. The main reason is that the measured values are contaminated with measurement errors. Anyway, let us try the relation $e + \kappa F = l$ for the measurements. We get

$$e + \kappa 1 = 7.97,$$
$$e + \kappa 2 = 10.2,$$
$$e + \kappa 3 = 14.2,$$
$$e + \kappa 4 = 16.0,$$
$$e + \kappa 5 = 21.2,$$

or, in matrix–vector form

$$\begin{pmatrix} 1 & 1 \\ 1 & 2 \\ 1 & 3 \\ 1 & 4 \\ 1 & 5 \end{pmatrix} \begin{pmatrix} e \\ \kappa \end{pmatrix} = \begin{pmatrix} 7.97 \\ 10.2 \\ 14.2 \\ 16.0 \\ 21.2 \end{pmatrix}.$$

In the example, we derived an **overdetermined linear system of equations**, i.e., a system with more equations than unknowns. In the absence of mesurement errors, it would have been possible to compute the equation of the straight line by interpolating in two points, but here, if we choose the the measured values for $F = 3$ and $F = 4$, the slope of this line will differ very much from the slope of the line we obtain with the measured values for $F = 2$ and $F = 4$. The purpose of using more equations than unknowns is to *reduce the influence of measurement errors*.

More generally, assume that we have an overdetermined system

$$Ax = b,$$

where A is an $m \times n$ matrix, $m > n$, x is a vector in R^n and b is a vector in R^m. As in Section 8.1, we can now partition the matrix into column vectors

$$A = (a_{.1}\, a_{.2} \ldots a_{.n}), \qquad a_{.j} = \begin{pmatrix} a_{1j} \\ a_{2j} \\ \vdots \\ a_{mj} \end{pmatrix},$$

and then we can write the overdetermined system $Ax = b$ in the form

$$x_1 a_{.1} + x_2 a_{.2} + \ldots + x_n a_{.n} = b.$$

The problem of solving the system of equations is equivalent to determining a linear combination of the column vectors so that it is equal to the right hand side b. In general, this is impossible: we have n vectors, $a_{.1}, a_{.2}, \ldots, a_{.n}$, and, since $m > n$, they cannot be a basis in R^m, which is a space of dimension m.

Example 8.13.3 Assume that $m = 3$ and $n = 2$. We then have two vectors $a_{.1}$ and $a_{.2}$ in R^3, and we want to solve an overdetermined system

$$x_1 a_{.1} + x_2 a_{.2} = b.$$

If the vectors are linearly independent they span a plane in R^3. We cannot presuppose that the vector b lies in this plane, see Figure 8.13.4.

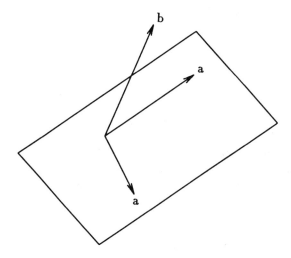

Figure 8.13.4

As we cannot solve the overdetermined system $Ax = b$, we shall have to be satisfied if we can determine a vector x, that makes the **residual vector**

$$r = b - Ax$$

small. Choosing to measure the smallness of the residual vector with the Euclidean vector norm (see Section 8.10)

$$\|r\|_2 = \left(\sum_{i=1}^{m} r_i^2 \right)^{1/2},$$

we then seek the least squares solution of the system: x is determined so that the sum of squares of the components of the residual vector is minimal. Equivalently, $\|r\|_2$ is minimized.

Given an overdetermined system of equations $Ax = b$, where A is an $m \times n$ matrix, $m > n$. In the **least squares method,** we minimize the Euclidean norm of the residual vector, i.e., x is determined as the solution of the minimization problem

$$\min_{x} \|b - Ax\|_2.$$

In the special case $m = 3$, $n = 2$, we immediately see geometrically how we can solve the least squares problem.

Example 8.13.5   According to Example 8.13.3 and the formulation of the least squares problem, we shall determine a linear combination of the vectors $a_{.1}$ and $a_{.2}$ so that the residual vector becomes as short as possible. Since all linear combinations of the two vectors lie in the plane in the figure, the length of the residual vector r is minimal if r is a normal to the plane (see Figure 8.13.6).

The requirement that r is normal to the plane is the same as r being normal to the vectors that span the plane, i.e.,

$$a_{.1}^T r = a_{.2}^T r = 0.$$

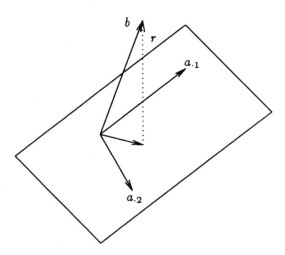

Figure 8.13.6   Solution of the least squares problem.

In the general case, the solution is obtained analogously.

Theorem 8.13.7   If the column vectors of A are linearly independent, then the matrix A^TA is positive definite, and the least squares problem

$$\min_x \|Ax - b\|_2$$

has a unique solution, which is obtained by solving the **normal equations**

$$A^TAx = A^Tb.$$

Before we prove the theorem, we note that the normal equations can be written

$$A^T(b - Ax) = A^Tr = 0.$$

Denote the solution of the normal equations and the corresponding residual vector $\hat{x}$ and $\hat{r} = b - A\hat{x}$, respectively. If we partition A in column vectors, we get

$$A^T\hat{r} = \begin{pmatrix} a_{\cdot 1}^T \\ a_{\cdot 2}^T \\ \vdots \\ a_{\cdot n}^T \end{pmatrix} \hat{r} = 0,$$

i.e., the column vectors of A are orthogonal to $\hat{r}$, which thus satisfies the "geometric" optimality condition.

Proof.   We first show that A^TA is positive definite. Let x be an arbitrary nonzero vector. Then, from the definition of linear independence, we have $Ax \neq 0$. With $y = Ax$, we then have

$$x^TA^TAx = y^Ty = \sum_{i=1}^{n} y_i^2 > 0,$$

which is equivalent to A^TA being positive definite. Therefore, A^TA is nonsingular, and the normal equations have a unique solution, which we denote $\hat{x}$.

Then, we show that $\hat{x}$ is the solution of the least squares problem, i.e., $\|\hat{r}\|_2 \leq \|r\|_2$ for all $r = b - Ax$. We can write

$$r = b - A\hat{x} + A(\hat{x} - x) = \hat{r} + A(\hat{x} - x),$$

and

$$\|r\|_2^2 = r^T r = (\hat{r} + A(\hat{x} - x))^T (\hat{r} + A(\hat{x} - x))$$
$$= \hat{r}^T \hat{r} + \hat{r}^T A(\hat{x} - x) + (\hat{x} - x)^T A^T \hat{r} + (\hat{x} - x)^T A^T A(\hat{x} - x).$$

Since $A^T \hat{r} = 0$, the two terms in the middle are equal to zero, and we get

$$\|r\|_2^2 = \hat{r}^T \hat{r} + (\hat{x} - x)^T A^T A(\hat{x} - x) = \|\hat{r}\|_2^2 + \|A(\hat{x} - x)\|_2^2 \geq \|\hat{r}\|_2^2,$$

which was to be proved.   ∎

We now solve the overdetermined system in Example 8.13.1 in the sense of the least squares method.

Example 8.13.8   We have the least squares problem min $\|Ax - b\|_2$, where

$$A = \begin{pmatrix} 1 & 1 \\ 1 & 2 \\ 1 & 3 \\ 1 & 4 \\ 1 & 5 \end{pmatrix},$$

and

$$b = \begin{pmatrix} 7.97 \\ 10.2 \\ 14.2 \\ 16.0 \\ 21.2 \end{pmatrix}.$$

The normal equations are

$$\begin{pmatrix} 5 & 15 \\ 15 & 55 \end{pmatrix} x = \begin{pmatrix} 69.57 \\ 240.97 \end{pmatrix},$$

and they have the solution

$$x = \begin{pmatrix} 4.236 \\ 3.226 \end{pmatrix}.$$

The straight line that best approximates the measured values in the sense of the least squares method is illustrated in Figure 8.13.9.

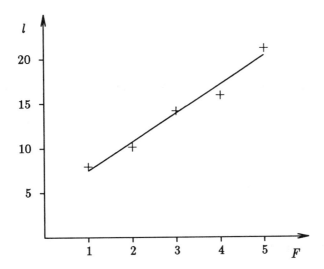

Figure 8.13.9   The least squares solution.

Returning to the notation of the original example, we have gotten $\kappa = 3.226$, and the spring constant $k = \kappa^{-1} \approx 0.310$.

In the previous example, we fitted a straight line to given data. The equation of the straight line can be written in many ways. We now give an example that shows that the normal equations can become very ill-conditioned, if the representation of the approximating function is badly chosen.

Example 8.13.10   Assume that we shall fit a straight line to the following data:

t	$f(t)$
998	3.765
999	4.198
1000	5.123
1001	5.888
1002	6.184

If we write the equation of the straight line in the form $f^* = c_0 + c_1 t$, then we get a least squares problem $\min \|Ac - b\|_2$ with

$$
A = \begin{pmatrix} 1 & 998 \\ 1 & 999 \\ 1 & 1000 \\ 1 & 1001 \\ 1 & 1002 \end{pmatrix}, \qquad b = \begin{pmatrix} 3.765 \\ 4.198 \\ 5.123 \\ 5.888 \\ 6.184 \end{pmatrix}.
$$

The matrix in the normal equations becomes

$$
C = A^T A = \begin{pmatrix} 5 & 5000 \\ 5000 & 5000010 \end{pmatrix},
$$

and its inverse is

$$
C^{-1} = \begin{pmatrix} 100000 & -100 \\ -100 & 0.1 \end{pmatrix}.
$$

We see that the normal equations are very ill-conditioned, and the condition number of C is

$$
\kappa(C) = \|C\|_\infty \cdot \|C^{-1}\|_\infty \approx 5 \cdot 10^6 \cdot 10^5 = 5 \cdot 10^{11},
$$

which makes the system impossible to solve on many computers.

If instead we choose the representation $f^* = b_0 + b_1(t - 1000)$, we get

$$
A = \begin{pmatrix} 1 & -2 \\ 1 & -1 \\ 1 & 0 \\ 1 & 1 \\ 1 & 2 \end{pmatrix},
$$

and the matrix in the normal equations becomes

$$
A^T A = \begin{pmatrix} 5 & 0 \\ 0 & 10 \end{pmatrix},
$$

with the condition number $\kappa(A^T A) = 2$.

8.14 High Performance Computers

Since linear algebra computations are so common, it is of primary importance that the algorithms are well implemented. In particular, it is necessary that large-scale problems execute efficiently on computers with vector instructions and/or parallel architecture. In this section, we discuss a few aspects of linear algebra computations on high performance computers, without going into too much detail. We first discuss a few hardware- and software-related factors that influence the design of efficient algorithms for fast computers.

As memory access patterns are important on high performance computers, we need to specify here how matrices are stored in primary memory. In Fortran, which is the most commonly used language for scientific computations, matrices are stored in column major order. E.g., a 3×3 matrix A is stored

a_{11}	a_{21}	a_{31}	a_{12}	a_{22}	a_{32}	a_{13}	a_{23}	a_{33}

Address: k $k + 8$

There are programming languages where matrices are stored rowwise, e.g., Pascal. In the sequel, we presuppose the Fortran storage convention.

Very often, the primary memory of high performance computers is interleaved: it consists of separate banks, which can operate independently of each other. Reading or writing a word from a bank takes a certain number of clock cycles. The consecutive elements of a vector are stored in consecutive banks, see Figure 8.14.1.

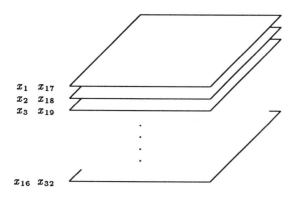

x_1 x_{17}
x_2 x_{18}
x_3 x_{19}

x_{16} x_{32}

Figure 8.14.1   The storage of a vector in interleaved memory with 16 banks.

Since the different banks can operate independently, it is possible to load consecutive elements of a vector from memory to vector registers (or the other way around), so that one word is delivered each clock cycle. On the other hand, if one loads non-consecutive elements of the vector, it may happen that when a word is requested from a memory bank, the bank has not finished processing the previous request. This is called a **memory bank conflict**.

Memory bank conflicts may occur whenever non-unit **stride** references to a vector are made (the statement "**for** $i := 1$ **step 2 to** n **do** $x_i :=$ ..." is a stride 2 reference to the vector x). Note that since matrices are stored in column major order, referencing a matrix rowwise is equivalent to referencing a vector with non-unit stride.

Thus, if linear algebra algorithms are organized so that they reference matrices columnwise they **vectorize** well: *the primary memory can deliver operands fast enough to keep the pipelined arithmetic of the functional units working at full speed.* The algorithm can be executed by **vector instructions**, cf. Section 2.10.

The observation that it is better to organize matrix computations columnwise was used in the construction of **LINPACK**, which is a subroutine library for linear systems of equations and least squares problems. This library is based on a set of computational kernels, the **Blas** (Basic Linear Algebra Subprograms; actually, for reasons that will be commented upon later, they are nowadays called level 1 Blas). We illustrate the use of Blas kernels for the algorithm for Gaussian elimination given in Section 8.6. For simplicity, we omit the pivoting here.

The basic Gaussian elimination algorithm is given by the following code.

> **for** $k := 1$ **to** $n - 1$ **do**
>   **for** $i := k + 1$ **to** n **do** $a_{ik} := a_{ik}/a_{kk}$;
>   **for all** (i, j), $k + 1 \leq i, j \leq n$ **do**
> $a_{ij} := a_{ij} - a_{ik} * a_{kj}$;

In the presentation in Sections 8.3–8.6, we intentionally did not specify the order in which the operations of the (i, j)-loop were to be performed, since the order is irrelevant from a mathematical point of view. But, the discussion in this section shows that, for many computers, it is better to write the inner loops in the following order.

> **for** $k := 1$ **to** $n - 1$ **do**
>   **for** $i := k + 1$ **to** n **do** $a_{ik} := a_{ik}/a_{kk}$;
>   **for** $j := k + 1$ **to** n **do**
>   **for** $i := k + 1$ **to** n **do**
> $a_{ij} := a_{ij} - a_{kj} * a_{ik}$;

Note that, in the innermost loop, the variable a_{kj} is constant, and the operation of the i-loop is of the form

$$y := y + \alpha x,$$

where x and y are vectors and α is a scalar. If we define a procedure saxpy,

```
procedure saxpy(m, n, α, x, y);
    for i := m to n do
        y_i := y_i + α * x_i;
```

and a procedure scale

```
procedure scale(m, n, β, x);
    for i := m to n do
        x_i := β * x_i;
```

then we can write the code for Gaussian elimination

```
for k := 1 to n − 1 do
    scale(k + 1, n, 1/a_{kk}, a_{.k});
    for j := k + 1 to n do
        saxpy(k + 1, n, −a_{kj}, a_{.k}, a_{.j});
```

There are also Blas routines for scalar products, for computing the norm of a vector, and for pivoting operations. What we have described here are only sketches of Blas routines. The actual Blas routines are defined in Fortran, and they have options for unit stride and non-unit stride, there are double precision versions, and so on. On many computers, the Blas routines are programmed in assembler language, and optimized for the architecture of that specific computer.

Another important concept in high performance computers is **memory hierarchy**. Due to the cost of manufacturing very fast memory hardware, computer designers often must compromise between memory speed and size. Many modern high performance computers, therefore, have a memory hierarchy, see Figure 8.14.2. (We remark that some computers have vector registers and no cache memory, while others have cache and no vector registers. There are computers that have both.)

Obviously, it is desirable to *perform as many floating point operations as possible for every floating point variable that is transferred from primary memory to the registers or cache memory.*

Note that the Blas operations that we have sketched above are vector operations; they access $O(n)$ data (if the vectors are in R^n), and perform $O(n)$ floating point operations. If, instead, we can organize Gaussian elimination in terms of matrix blocks, then we can get more operations per memory reference.

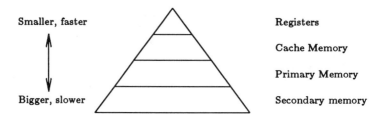

Figure 8.14.2   Memory hierarchy.

Consider the $n \times n$ matrix

$$A = \begin{pmatrix} A_{11} & A_{12} & A_{13} \\ A_{21} & A_{22} & A_{33} \\ A_{31} & A_{32} & A_{33} \end{pmatrix},$$

where, for simplicity, we assume that the blocks are $\alpha \times \alpha$, $\alpha = n/3$. Further, we disregard pivoting here. Then, we write the first step of **block Gaussian elimination**:

$$\begin{pmatrix} A_{11} & A_{12} & A_{13} \\ A_{21} & A_{22} & A_{33} \\ A_{31} & A_{32} & A_{33} \end{pmatrix} = \begin{pmatrix} L_{11} & 0 & 0 \\ L_{21} & I & 0 \\ L_{31} & 0 & I \end{pmatrix} \begin{pmatrix} U_{11} & U_{12} & U_{13} \\ 0 & A'_{22} & A'_{23} \\ 0 & A'_{32} & A'_{33} \end{pmatrix}.$$

Identifying blocks, we get

1) $L_{11}U_{11} = A_{11}$,
2) $L_{11}U_{12} = A_{12}$, $L_{11}U_{13} = A_{13}$,
3) $L_{21}U_{11} = A_{21}$, $L_{31}U_{11} = A_{31}$,
4) $A'_{22} = A_{22} - L_{21}U_{12}$, $A'_{23} = A_{23} - L_{21}U_{13}$,
    $A'_{32} = A_{32} - L_{31}U_{12}$, $A'_{33} = A_{33} - L_{31}U_{13}$.

Thus, with an obvious generalization to the case where we have N^2 blocks (instead of nine), each of dimension $\alpha = n/N$, we can write the algorithm for block Gaussian elimination:

> **for** $k := 1$ **to** N **do**
> $\qquad [L_{kk}, U_{kk}] := lu(A_{kk}, \alpha);$
> $\qquad$ **for** $j := k+1$ **to** N **do** $U_{kj} := L_{kk}^{-1} A_{kj};$
> $\qquad$ **for** $i := k+1$ **to** N **do** $L_{ik} := A_{ik} U_{kk}^{-1};$
> $\qquad$ **for all** (i, j), $k+1 \leq i, j \leq N$ **do**
> $\qquad\qquad A_{ij} := A_{ij} - L_{ik} U_{kj};$

Note that the first assignment is a conventional Gaussian elimination. The next two loops are really solutions of triangular system with multiple right hand sides. The (i, j) loop consists of matrix multiplications and additions. Thus, the computation of any of the blocks in L, U and the reduced matrix requires $O(\alpha^3)$ floating point operations, and, if the cache memory can hold three matrix blocks, then $O(\alpha^2)$ data must be moved to the cache for each matrix operation. Thus, the floating point operations to memory reference ratio is $O(\alpha)$, in contrast to the ratio $O(1)$ in the conventional algorithm based on vector–vector operations.

The level 3 Blas library contains subroutines for matrix–matrix operations of the type illustrated above (level 2 Blas performs matrix–vector operations; we do not mention this further). Typical of level 3 Blas routines is what can be called the **surface to volume effect**: there are $O(\alpha^2)$ memory references and $O(\alpha^3)$ floating point operations.

Another important aspect here is **parallelism**: the computations of the different iterations of each loop in the algorithm above are completely independent. They can therefore be performed in parallel by different processors in a parallel computer. We express this parallelism using the construct **parallel — do**:

$$\textbf{parallel } k+1 \leq j \leq N \textbf{ do } U_{kj} := L_{kk}^{-1} A_{kj};$$

In practice, the matrices L and U overwrite A. Let us assume that we have a conventional LU decomposition procedure

$$\textbf{procedure } \text{lu}(k, A);$$

which computes the factors L and U of the $k \times k$ matrix A, and stores them below and above the main diagonal of A. Further, the procedure

$$\textbf{procedure } \text{trilsolve}(k, A, B);$$

solves the lower triangular system $LX = B$, where the $k \times k$ matrix L is stored below the main diagonal of A. The result overwrites B, which is also assumed to be $k \times k$. Similarly, the procedure

procedure triusolve(k, A, B);

solves the upper triangular system $UX = B$, where U is stored above the main diagonal of A. Finally,

procedure mmul(k, A, B, C);

computes $C := C + AB$. With these procedures, we can write a **parallel block LU decomposition algorithm**:

> **for** $k := 1$ **to** N **do**
>   lu(α, A_{kk});
>   **parallel** $k + 1 \leq j \leq N$ **do** trilsolve(α, A_{kk}, A_{kj});
>   **parallel** $k + 1 \leq i \leq N$ **do** triusolve(α, A_{kk}, A_{ik});
>   **parallel** $k + 1 \leq i, j \leq N$ **do** mmul($\alpha, -A_{ik}, A_{kj}, A_{ij}$);

The level 3 Blas routines can be optimized for a certain computer architecture and coded in assembler language. The Fortran programs based on these routines are machine independent, and will perform very well over a wide range of computers.

Exercises

1. Solve the system of equations

$$
\begin{aligned}
0.5x_1 \quad\quad\quad + x_3 &= 1, \\
x_1 + 2x_2 - x_3 &= 0, \\
x_1 \quad\quad\quad + x_3 &= 0,
\end{aligned}
$$

using Gaussian elimination and partial pivoting.

2. Determine the permutation matrix P such that

$$
P \begin{pmatrix} x_1 \\ x_2 \\ x_3 \\ x_4 \\ x_5 \end{pmatrix} = \begin{pmatrix} x_4 \\ x_1 \\ x_3 \\ x_2 \\ x_5 \end{pmatrix}.
$$

3. a) Solve the system $Ax = b$ with Gaussian elimination and partial pivoting, for

$$A = \begin{pmatrix} 0.8 & 1.4 & 3 \\ 0.6 & 0.9 & 2.8 \\ 2 & 1 & 0 \end{pmatrix}, \qquad b = \begin{pmatrix} 12.6 \\ 10.8 \\ 4 \end{pmatrix}.$$

b) Determine P, L and U in the LU decomposition $PA = LU$.

4. a) Compute a LU decomposition of the matrix

$$A = \begin{pmatrix} 1.4 & 1.42 & 6.5 \\ 2 & 1 & 1 \\ 0.4 & 1.4 & 3.2 \end{pmatrix},$$

using Gaussian elimination and partial pivoting.
b) Use the decomposition from a) to solve the system $Ax = b$, where $b = (39.58, 11, 22)^T$.

5. When solving systems of equations of the type

$$\begin{pmatrix} 10 & -1 & 1 & -1 & 1 \\ -1 & 5 & 0 & 0 & 0 \\ 1 & 0 & 2 & 0 & 0 \\ -1 & 0 & 0 & 5 & 0 \\ 1 & 0 & 0 & 0 & 2 \end{pmatrix} x = \begin{pmatrix} 42.8 \\ 1.5 \\ 9.1 \\ 12.5 \\ 4.7 \end{pmatrix},$$

it is useful to exchange rows 1 and 5, and columns 1 and 5 (Why?). Solve the system after having permuted the columns, and determine the LU decomposition of the permuted matrix.

6. a) Compute the LDL^T decomposition of the matrix

$$A = \begin{pmatrix} 4 & 2 & 0 & 0 \\ 2 & 5 & 2 & 0 \\ 0 & 2 & 5 & 2 \\ 0 & 0 & 2 & 5 \end{pmatrix}.$$

b) Compute the Cholesky decomposition.

7. Solve the least squares problem $\min \|Ax - b\|$, where

$$A = \begin{pmatrix} 1 & -3 & 1 \\ 3 & 1 & -11 \\ 1 & -2 & -1 \\ 2 & 1 & 1 \end{pmatrix}, \qquad b = \begin{pmatrix} 1 \\ 1 \\ 1 \\ 2 \end{pmatrix}.$$

8. Fit a second degree polynomial to the following measured values:

x	1	2	3	6	8
$f(x)$	2.2	1.8	1.7	1.3	0.9

using the least squares method.

9. Let B be a positive definite matrix. Show that $\|x\|_B = (x^T B x)^{1/2}$ is a vector norm.

10. Let $\|\cdot\|$ be a matrix norm corresponding to a vector norm. Show that if D is a diagonal matrix, $D = \mathrm{diag}(d_1, d_2, \ldots, d_n)$, then $\|D\| = \max|d_i|$.

11. Let $P = I - 2ww^T$, where $\|w\|_2 = 1$.
a) Show that P is orthogonal, i.e., $P^T P = I$.
b) Show that $\|Px\|_2 = \|x\|_2$.

12. Given the system of equations $Ax = b$, where

$$A = \begin{pmatrix} 0.5 & 0.4 \\ 0.3 & 0.25 \end{pmatrix},$$

and where we have only an approximation $\bar{b} = (0.200, 1.000)^T$ of the right hand side. Assuming that the approximation is correctly rounded to three decimals, give an estimate for the uncertainty in the solution, $\|\delta x\|_\infty / \|x\|_\infty$.

13. Given

$$A = \begin{pmatrix} 10^{-3} & 1 & -1 \\ 1 & 1 & 1 \\ -1 & 1 & 1 \end{pmatrix}, \qquad b = \begin{pmatrix} 0.117 \\ 0.352 \\ 0.561 \end{pmatrix}.$$

a) Compute an LU decomposition of A using Gaussian elimination and partial pivoting.
b) Use the decomposition from a) to compute A^{-1}.
c) Assume that the components of b are correctly rounded. Give an upper bound for the relative uncertainty in the solution of $Ax = b$ (use maximum norm).

14. a) Compute an LU decomposition of the matrix

$$A = \begin{pmatrix} 20 & 3 & 4 \\ 3 & 40 & 5 \\ 4 & 5 & 60 \end{pmatrix},$$

using the floating point system $(10, 1, -9, 9)$.
b) The system $Ax = b$, where $b = (15, -360, 420)^T$, has the approximate solution $x^{(1)} = (0.65, -9.8, 7.8)^T$. Compute a better approximation using one step of iterative refinement.

References

As is apparent from the name, Gaussian elimination is a classical algorithm, which has been used for a very long time for the solution of small systems with very simple calculating tools. When in the late 1940s it became possible to solve systems with relatively many unknowns, there were many who believed that the method was unstable in the sense that rounding errors would grow catastrophically and eventually dominate the solution. These fears turned out to be exaggerated, and the method can be used for very large systems. The first rounding error analysis for Gaussian elimination was developed by J. H. Wilkinson in the early 1950s and is described in the book

J. H. Wilkinson, *Rounding Errors in Algebraic Processes*, Prentice–Hall, Englewood Cliffs, New Jersey, 1963.

Wilkinson summarized the "state of the art" in numerical linear algebra in the middle of the 1960s:

J. H. Wilkinson, *The Algebraic Eigenvalue Problem*, Clarendon Press, Oxford, 1965.

This book had a great influence on the development of software for the solution of linear systems of equations, where the matrix is dense or has band structure. The following is a documentation of a program library for linear algebra

J. J. Dongarra et al., *LINPACK User's Guide*, SIAM, Philadelphia, 1979.

The "state of the art" in numerical linear algebra around 1990 is presented in

G. H. Golub and C. F. Van Loan, *Matrix Computations*, Second Edition, Johns Hopkins Press, Baltimore, Maryland, 1989.

This book also gives an excellent introduction to linear algebra algorithms for vector and parallel computers. Level 3 Blas subroutines are discussed in

J. J. Dongarra, J. Du Croz, I. Duff and S. Hammarling, A Set of Level 3 Basic Linear Algebra Subprograms, Preprint No.1, Mathematics and Computer Science Division, Argonne National Laboratory, Argonne, Illinois, August 1988.

9 Approximation

9.1 Introduction

In Chapter 5, we approximated a given function by an interpolating polynomial, i.e., a polynomial that agrees exactly with the function at certain points. In many cases, however, it is not appropriate to let the approximating function be an interpolating polynomial.

Assume, for instance, that the function f describes the relation between two physical entities x and $f(x)$. By measurement, one has determined $f_i = f(x_i) + \epsilon_i$, $i = 0, 1, \ldots, M$, where the errors ϵ_i are unknown. (We assume that the x_i are exact.) Unless the errors are very small, it is unreasonable to describe f by an interpolating polynomial through these measurement points. If, instead, the polynomial is determined using the method of least squares, we reduce the influence of the measurement errors. We shall give a detailed description of this method below (cf. also Example 8.13.1.)

The situation is different when a function given by an analytic expression, e.g., e^x, is to be approximated by a polynomial in order to be evaluated on a computer. Here, we are interested in the polynomial of lowest possible degree that approximates the given function on an interval with a maximal error less than a certain tolerance.

In both cases, there is a given function f that is to be approximated by a simpler function f^*. In the sequel, f^* will mostly be a polynomial, since polynomials are easy to evaluate, differentiate, and integrate. But these are not the only requirements. The approximating function should also have a similar behaviour as the given function. Therefore, in many cases it is more natural to approximate by other functions, e.g., trigonometric polynomials, exponential functions, or rational functions. For instance, it can be shown that, for $-1 \leq x \leq 1$, the function $f(x) = \tan x$ can be approximated with a maximal error less than $2 \cdot 10^{-5}$ by a suitable choice of the three parameters a_0, a_1 and a_2 in the rational function

$$f^*(x) = x \left(a_0 + \frac{a_1}{a_2 - x^2} \right).$$

In order to obtain the same accuracy with a polynomial, it is necessary to choose the degree to be at least nine.

Sometimes, a simple trick can be used to reduce a general approximation problem to a problem involving the approximation by a polynomial. Assume, e.g., that $f(x) = \cot x$ shall be approximated. Since

$$\cot x = \frac{1}{x} \left(1 - \frac{x^2}{3} - \frac{x^4}{45} - \cdots \right) \qquad \text{for} \quad x^2 < \pi^2,$$

it is natural to determine a polynomial of the type $\sum_{k=0}^{N} a_k x^{2k}$, which approximates $x \cot x$.

Another possibility to get good accuracy in the approximation by polynomials is to approximate by *different* polynomials in different parts of the interval (cf Section 5.9).

What shall we mean then by a "good" approximation? In order that f^* be a good approximation to a given function f, we require the error function $f - f^*$ to be small in some sense. Assume, for instance, that $f(x) = e^x$ shall be approximated by a straight line $f^*(x) = c_0 + c_1 x$ on the interval $0 \leq x \leq 1$. The parameters c_0 and c_1 can then be determined, e.g., using the **method of least squares**, i.e., so that

$$\int_0^1 (f(x) - f^*(x))^2 \, dx = \int_0^1 (e^x - (c_0 + c_1 x))^2 \, dx$$

is minimized. This gives (see Example 9.3.4)

$$f_1^*(x) = (4e - 10) + 6(3 - e)x \approx 0.8731 + 1.6903x.$$

Alternatively, c_0 and c_1 can be determined so that we get the **best approximation in Chebyshev norm**, i.e., so that

$$\max_{0 \leq x \leq 1} |f(x) - f^*(x)| = \max_{0 \leq x \leq 1} |e^x - (c_0 + c_1 x)|$$

is minimized. This requirement gives a different function (see Example 9.7.4):

$$f_2^*(x) = \frac{1}{2}(e - (e - 1)\log(e - 1)) + (e - 1)x \approx 0.8941 + 1.7183x.$$

The error function $e^x - f_2^*$ has its maximal value equal to $1 - c_0 \approx 0.1059$, and this value is attained with alternating sign at $x_0 = 0$, $x_1 = \log(e-1) \approx 0.54$, and $x_2 = 1$ (see Figure 9.1.1). The error function $e^x - f_1^*$ has relatively large values at the end points of the interval: 0.1269 for $x = 0$, and 0.1548 for $x = 1$.

For functions other than e^x, the difference between the least squares approximation and the best approximation in Chebyshev norm may be considerably larger (cf. Exercise 9).

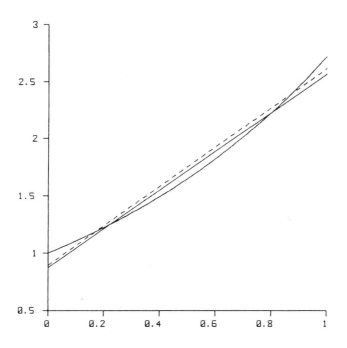

Figure 9.1.1   Different approximations to e^x: f_1^* (solid), f_2^* (dashed).

9.2   Some Important Concepts

We want to formulate the approximation problem in quite general terms, so that the same terminology can be used to describe the *continuous case*, when f is a continuous function defined on an interval $[a, b]$, as well as the *discrete case*, when f is known only at a number of points $x_0, x_1, \ldots, x_M$.

In both cases, f is an element of a **linear space**, and the approximating function f^* is chosen from a **subspace**. We therefore start by reminding the reader of the definition of these two concepts.

Definition 9.2.1   A space L is said to be **linear**, if for arbitrary elements f and g in L, and an arbitrary real number α, αf and $f + g$ also belong to L.

Definition 9.2.2   A nonempty subset U of a linear space L is said to be a **subspace** if, for arbitrary elements f_1 and f_2 in U, and an arbitrary real number α, αf_1 and $f_1 + f_2$ also belong to U.

The linear spaces that we are going to use are the space of all functions continuous on the interval $[a, b]$ (this space is denoted $C[a, b]$), and the space of all real valued vectors with n elements (denoted R^n). In the continuous case, we have $f \in C[a, b]$. In the discrete case, we shall approximate the vector

$$\text{tab } f = (f(x_0), f(x_1), \ldots, f(x_M))^T,$$

which is an element in R^{M+1}.

The approximating function f^* shall be of the form

$$f^*(x) = c_0\varphi_0(x) + c_1\varphi_1(x) + \ldots + c_N\varphi_N(x),$$

where $\varphi_0, \varphi_1, \ldots, \varphi_N$ are given continuous functions. The function f^* and the vector tab f^* belong to subspaces of $C[a, b]$ and R^{M+1}, respectively.

When approximating by polynomials, we may choose $\varphi_i(x) = x^i$, $i = 0, 1, \ldots, N$. However, in Example 8.13.10, we saw that this may be a bad choice. Later, we shall show that it is better to express f^* as a linear combination of **orthogonal** polynomials. It is not necessary to choose the functions $\varphi_i(x)$ to be polynomials, however. The theory is valid for approximation by a linear combination of a finite number of arbitrarily chosen continuous functions.

To measure the magnitude of the error function, we need to define the concept of a norm.

> **Definition 9.2.3**   Let f and g be elements in a linear space L. A
> **norm** $\| \cdot \|$ is a mapping $L \to R$, with the properties
>
> $$\|f\| \geq 0 \quad \text{for all } f,$$
> $$\|f\| = 0 \Leftrightarrow f = 0,$$
> $$\|\alpha f\| = |\alpha| \cdot \|f\| \quad \text{for an arbitrary real number,}$$
> $$\|f + g\| \leq \|f\| + \|g\|. \quad \text{(the triangle inequality)}$$

We use $\|f - g\|$ to measure the distance between two elements f and g in L, and get

> **The approximation problem.**   Let U be a subspace of the normed
> linear space L. Given $f \in L$, determine $f^* \in U$ such that
>
> $$\|f - f^*\| = \min_{g \in U} \|f - g\|.$$

The most common definitions of the norm of a function $f \in C[a,b]$ are:

$$\|f\|_2 = \sqrt{\int_a^b f(x)^2 \, dx}, \qquad \textbf{Euclidean norm,}$$

$$\|f\|_\infty = \max_{a \leq x \leq b} |f(x)|, \qquad \textbf{maximum norm or Chebyshev norm.}$$

Both these norms are special cases of the L_p-**norm**

$$\|f\|_p = \left(\int_a^b |f(x)|^p \, dx \right)^{1/p}.$$

It can be shown that, for $p \geq 1$, this expression defines a norm, i.e., that it has the properties in Definition 9.2.3.

For functions defined on a **net** $G = \{x_i\}_{i=0}^M$, the corresponding norm is defined as

$$\|f\|_{p,G} = \left(\sum_{i=0}^M |f(x_i)|^p \right)^{1/p}.$$

This norm is said to be a **seminorm**, since it does not satisfy all the requirements of Definition 9.2.3. The seminorm of a function f can be equal to zero without f being identically zero; it is enough that it be zero on the net G.

The definitions of the norm can be generalized by introducing a so-called weight function. This is a function w such that $w(x) > 0$ for all $a < x < b$. (w need not be defined for $x = a$ and $x = b$.) In the case of the seminorm, positive weights w_i are introduced. We get:

$$\|f\|_{p,w} = (\int_a^b w(x)|f(x)|^p \, dx)^{1/p},$$

$$\|f\|_{p,G,w} = (\sum_{i=0}^M w_i|f(x_i)|^p)^{1/p}.$$

For the *Euclidean* seminorm, we have

$$\|f\|_{2,G} = (\sum_{i=0}^M |f(x_i)|^2)^{1/2} = ((\text{tab } f)^T \text{tab } f)^{1/2} = \|\text{tab } f\|_2.$$

Thus, the norm of f is the Euclidean length of the vector tab f. Now we introduce a scalar product in the space L.

Definition 9.2.4   The **scalar product** (f, g) of f and $g \in L$ is defined as

$$(f, g) = \begin{cases} \int_a^b w(x)f(x)g(x) \, dx, & \text{(the continuous case)} \\ \\ \sum_{i=0}^M w_i f(x_i)g(x_i). & \text{(the discrete case)}. \end{cases}$$

For the Euclidean (semi-)norm, we immediately get

$$\|f\|_2 = \sqrt{(f, f)}.$$

With this scalar product, we can define orthogonal functions.

Definition 9.2.5   The functions f and g are said to be **orthogonal** if $(f, g) = 0$. A sequence of functions $\varphi_0, \varphi_1, \varphi_2, \dots$ is called an **orthogonal system** if $(\varphi_i, \varphi_j) = 0$ for $i \neq j$, and $(\varphi_i, \varphi_i) \neq 0$ for all i. If, in addition, $(\varphi_i, \varphi_i) = 1$ for all i, the sequence is called an **orthonormal system**.

Later, we shall show that the functions in an orthogonal system are linearly independent. We define

Definition 9.2.6   The functions $\varphi_0, \varphi_1, \dots, \varphi_N$ are said to be **linearly independent** if

$$\| \sum_{j=0}^{N} c_j \varphi_j \| = 0 \qquad \text{if and only if } c_j = 0, \quad j = 0, 1, \dots, N.$$

9.3   Euclidean Norm—Least Squares

We want to find an approximating function f^* that minimizes the Euclidean norm of the error function $f - f^*$.

In the **continuous case**, we shall determine f^* so that

$$\|f - f^*\|_{2,w} = \left(\int_a^b w(x)(f(x) - f^*(x))^2 \, dx \right)^{1/2}$$

is minimized. In the **discrete case**, we shall determine f^* so that

$$\|f - f^*\|_{2,G,w} = \left(\sum_{i=0}^{M} w_i (f(x_i) - f^*(x_i))^2 \right)^{1/2}$$

is minimized.

Let us first explain how the choice of w_i and w affects f^*. A large value of w_i means that the value of the error function at the point x_i is given special significance. For instance, it may be suitable to let w_i be inversely proportional to the estimated relative error in the measured value f_i. Then f^* is determined so that there is a particularly good agreement with f in the points that have been measured with good accuracy. Correspondingly, in the continuous case, by a suitable choice of the weight function w, we can force f^* to have better agreement with f in some part of $[a, b]$ than in the rest of the interval. Later, we shall see how this can be done in the context of Chebyshev approximation.

The following important theorem covers both the discrete and the continuous case.

Theorem 9.3.1   Assume that the functions $\varphi_0, \varphi_1, \ldots, \varphi_N$ are linearly independent. Then there is a uniquely determined function

$$f^* = \sum_{j=0}^{N} c_j^* \varphi_j,$$

such that

$$\|f - f^*\|_2 \leq \|f - g\|_2,$$

for all $g = \sum_{j=0}^{N} c_j \varphi_j$.
The function f^* is characterized by the **normal equations**

$$(f - f^*, \varphi_k) = 0, \qquad k = 0, 1, 2, \ldots, N.$$

We note that the normal equations can be written

$$(f^*, \varphi_k) = (f, \varphi_k), \qquad k = 0, 1, \ldots, N.$$

If we use the expression for f^* in the theorem we get

$$\sum_{j=0}^{N} c_j^* (\varphi_j, \varphi_k) = (f, \varphi_k), \qquad k = 0, 1, \ldots, N.$$

Thus, the coefficients c_j^*, $j = 0, 1, \ldots, N$, can be determined by solving a linear system of equations. Let us illustrate the theorem by a figure (for the case $N = 1$) before we prove it.

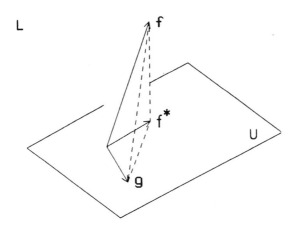

Figure 9.3.2   Geometric interpretation of the least squares problem.

The linear subspace U of L is spanned by $\varphi_0, \varphi_1, \ldots, \varphi_N$. According to the theorem, $f - f^*$ is orthogonal to $\varphi_0, \varphi_1, \ldots, \varphi_N$, and thereby to all vectors in U, i.e., f^* is the orthogonal projection of f onto U. The meaning of the theorem is simply that the orthogonal projection f^* of f onto U is the vector in U that has the least Euclidean distance to f. For the proof of the theorem, we are guided by the geometric interpretation. We first need a generalization of the Pythagorean law.

Theorem 9.3.3 If the functions f and g are orthogonal, then

$$\|f + g\|_2^2 = \|f\|_2^2 + \|g\|_2^2.$$

Proof.
$$\begin{aligned}
\|f + g\|_2^2 &= (f + g, f + g) \\
&= (f, f) + (f, g) + (g, f) + (g, g) \\
&= \|f\|_2^2 + \|g\|_2^2,
\end{aligned}$$

since $(f, g) = (g, f) = 0$ due to the orthogonality.   ∎

Proof (Theorem 9.3.1).   We first prove that the linear system of equations

$$\sum_{j=0}^{N} c_j(\varphi_j, \varphi_k) = (f, \varphi_k), \qquad k = 0, 1, \ldots, N,$$

has a unique solution, i.e., that the matrix of the system is non-singular.

Assume the opposite! Then the homogeneous system has a non-trivial solution, i.e., there are $c_0, c_1, \ldots, c_N$, not all equal to zero, such that

$$\sum_{j=0}^{N} c_j(\varphi_j, \varphi_k) = 0, \qquad k = 0, 1, \ldots, N.$$

But this means that $\varphi_0, \varphi_1, \ldots, \varphi_N$ are linearly dependent, since

$$\left\| \sum_{j=0}^{N} c_j \varphi_j \right\|_2^2 = \left(\sum_{k=0}^{N} c_k \varphi_k, \sum_{j=0}^{N} c_j \varphi_j \right)$$

$$= \sum_{k=0}^{N} c_k \left(\sum_{j=0}^{N} c_j(\varphi_j, \varphi_k) \right) = \sum_{k=0}^{N} c_k \cdot 0 = 0.$$

This contradicts the assumptions of the theorem.

Now, consider $f^* = \sum_{j=0}^{N} c_j^* \varphi_j$, and show that any function $g = \sum_{j=0}^{N} c_j \varphi_j$, with at least one $c_j \neq c_j^*$, has a greater distance to f than f^* (cf. Figure 9.3.2).

For arbitrary c_j, we have

$$f - \sum_{j=0}^{N} c_j \varphi_j = (f - f^*) + \left(f^* - \sum_{j=0}^{N} c_j \varphi_j \right)$$

$$= (f - f^*) + \sum_{j=0}^{N} (c_j^* - c_j) \varphi_j.$$

Since $(f - f^*, \varphi_j) = 0$ for $j = 0, 1, \ldots, N$, the following holds:

$$\left(f - f^*, \sum_{j=0}^{N} (c_j^* - c_j) \varphi_j \right) = 0.$$

Now the Pythagorean law gives

$$\|f - \sum_{j=0}^{N} c_j \varphi_j\|_2^2 = \|f - f^*\|_2^2 + \|\sum_{j=0}^{N}(c_j^* - c_j)\varphi_j\|_2^2,$$

i.e.,

$$\|f - \sum_{j=0}^{N} c_j \varphi_j\|_2^2 \geq \|f - f^*\|_2^2$$

for all c_j. Equality is obtained only if $c_j^* = c_j$, since the functions $\varphi_0, \varphi_1, \ldots, \varphi_N$ are linearly independent. ∎

Example 9.3.4   Determine the straight line $f^*(x) = c_0 + c_1 x$ that is the best approximation, in the least squares sense, to the function $f(x) = e^x$ on the interval $[0, 1]$.

Here $\varphi_0 = 1$, $\varphi_1 = x$.

$$(\varphi_0, \varphi_0) = \int_0^1 1 \, dx = 1,$$

$$(\varphi_0, \varphi_1) = (\varphi_1, \varphi_0) = \int_0^1 x \, dx = 1/2,$$

$$(\varphi_1, \varphi_1) = \int_0^1 x^2 \, dx = 1/3,$$

$$(f, \varphi_0) = \int_0^1 e^x \, dx = e - 1,$$

$$(f, \varphi_1) = \int_0^1 x e^x \, dx = 1.$$

The normal equations are

$$c_0 + \frac{1}{2}c_1 = e - 1,$$

$$\frac{1}{2}c_0 + \frac{1}{3}c_1 = 1,$$

and we get

$$c_0 = 4e - 10 \approx 0.8731,$$

$$c_1 = 6(3 - e) \approx 1.6903.$$

As the conclusion of this section, we consider the discrete case in detail in order to see the relation to the results in Section 8.13. We start by giving an alternative derivation of the normal equations.

The coefficients $c_0, c_1, \ldots, c_N$ are to be determined so that the square of the norm of the error function, i.e., the function

$$\phi(c_0, c_1, \ldots, c_N) = \sum_{i=0}^{M} w_i (f(x_i) - f^*(x_i))^2,$$

is minimized. ϕ is a continuously differentiable function of $c_0, c_1, \ldots, c_N$. At a minimal point of ϕ, we must have

$$\frac{\partial \phi}{\partial c_k} = 0, \qquad k = 0, 1, \ldots, N.$$

For each k, this can be written

$$-2 \sum_{i=0}^{M} w_i (f(x_i) - f^*(x_i)) \frac{\partial f^*(x_i)}{\partial c_k} = 0,$$

or

$$\sum_{i=0}^{M} w_i (f(x_i) - f^*(x_i)) \varphi_k(x_i) = 0,$$

or

$$c_0 \sum_{i=0}^{M} w_i \varphi_0(x_i) \varphi_k(x_i) + \ldots + c_N \sum_{i=0}^{M} w_i \varphi_n(x_i) \varphi_k(x_i) = \sum_{i=0}^{M} w_i \varphi_k(x_i) f(x_i).$$

Thus, we can determine the coefficients $c_0, c_1, \ldots, c_N$ from a linear system of equations. Using scalar products, this can be written

$$c_0 (\varphi_0, \varphi_k) + \ldots + c_i (\varphi_i, \varphi_k) + \ldots + c_N (\varphi_n, \varphi_k) = (f, \varphi_k),$$

or

$$(f^*, \varphi_k) = (f, \varphi_k), \qquad k = 0, 1, \ldots, N.$$

We recognize the normal equations from Theorem 9.3.1. Alternatively we can write the normal equations as in Theorem 8.13.7. Let us show that.

Consider the vector

$$\text{tab } f^* = \begin{pmatrix} c_0 \varphi_0(x_0) + c_1 \varphi_1(x_0) + \ldots + c_N \varphi_N(x_0) \\ \vdots \\ c_0 \varphi_0(x_M) + c_1 \varphi_1(x_M) + \ldots + c_N \varphi_N(x_M) \end{pmatrix},$$

and let A be the matrix with the columns $\text{tab } \varphi_i$, $i = 0, 1, \ldots, N$. Then we get

$$\text{tab } f^* = (\text{tab } \varphi_0 \quad \text{tab } \varphi_1 \quad \cdots \quad \text{tab } \varphi_N) \begin{pmatrix} c_0 \\ c_1 \\ \vdots \\ c_N \end{pmatrix} = Ac.$$

To determine f^* so that $\|f - f^*\|_{2,G}^2 = \|\text{tab } f - \text{tab } f^*\|_2^2$ is minimized is equivalent to solving the overdetermined system of equations

$$\text{tab } f^* = \text{tab } f, \quad \text{or} \quad Ac = b,$$

in the least squares sense. The normal equations of Theorem 8.13.7,

$$A^T A c = A^T b,$$

are identical to those we have derived, since

$$(A^T A)_{i,k} = (\text{tab } \varphi_i)^T \text{tab } \varphi_k = (\varphi_i, \varphi_k),$$
$$(A^T b)_k = (\text{tab } \varphi_k)^T \text{tab } f = (f, \varphi_k).$$

Further, according to Theorem 8.13.7, the normal equations have a unique solution if the column vectors of A are linearly independent. We shall show that this is satisfied if and only if the functions $\varphi_0, \varphi_1, \ldots, \varphi_N$ are linearly independent. In the discrete case, we have

$$\left\| \sum_{j=0}^{N} c_j \varphi_j \right\|_{2,G} = 0,$$

if and only if

$$\sum_{j=0}^{N} c_j \varphi_j(x_i) = 0 \quad \text{for } x_i \in G.$$

This can be written

$$c_0 \begin{pmatrix} \varphi_0(x_0) \\ \varphi_0(x_1) \\ \vdots \\ \varphi_0(x_M) \end{pmatrix} + \ldots + c_N \begin{pmatrix} \varphi_N(x_0) \\ \varphi_N(x_1) \\ \vdots \\ \varphi_N(x_M) \end{pmatrix} = 0,$$

or

$$\sum_{j=0}^{N} c_j \text{ tab } \varphi_j = 0.$$

Thus, we have

Theorem 9.3.5   The functions $\varphi_0, \varphi_1, \ldots, \varphi_N$ are linearly independent on the net $x_0, x_1, \ldots, x_M$ if and only if tab φ_j, $j = 0, 1, \ldots, N$,
are linearly independent vectors.

Corollary 9.3.6   There are at most $(M+1)$ linearly independent functions on a net with $(M+1)$ points.

Example 9.3.7   The functions $1, x, x^2, \ldots, x^N$ are linearly independent on *any* net with $(N+1)$ points. If they were not, then there
would exist constants $c_0, c_1, \ldots, c_N$, not all equal to zero, such that
$\sum_{j=0}^{N} c_j$ tab $\varphi_j = 0$. But, this means that

$$\sum_{j=0}^{N} c_j x_i^j = 0 \quad \text{for} \quad i = 0, 1, \ldots, N.$$

We would then have a polynomial of degree $\leq N$, not identically zero,
with $(N+1)$ zeros. This is impossible.

9.4   Orthogonal Polynomials

According to Theorem 9.3.1, we can determine the least squares approximation of a given function by solving the normal equations. In practice,
this system of equations may be ill-conditioned and numerical difficulties
occur.

Assume, e.g., that the function f shall be approximated by a polynomial f^* on the interval $[0, 1]$. If we represent the polynomial by a linear
combination of the functions $\varphi_j(x) = x^j$, $j = 0, 1, \ldots, N$, we get

$$(\varphi_i, \varphi_k) = \int_0^1 x^{i+k}\, dx = \frac{1}{i+k+1},$$

and the matrix of the normal equations is the Hilbert matrix

$$\begin{pmatrix} 1 & 1/2 & 1/3 & \ldots & 1/(N+1) \\ 1/2 & 1/3 & 1/4 & \ldots & 1/(N+2) \\ \vdots & & & & \vdots \\ 1/(N+1) & & \ldots & & 1/(2N+1) \end{pmatrix}.$$

This matrix is known to be ill-conditioned.

If, instead, we choose an orthogonal basis of the space of all polynomials of degree N, we avoid numerical difficulties of this kind. We shall show this. If the functions φ_j are chosen to be orthogonal, i.e., $(\varphi_j, \varphi_k) = 0$ for $j \neq k$, the normal equations become

$$c_k(\varphi_k, \varphi_k) = (f, \varphi_k), \qquad k = 0, 1, \ldots, N.$$

We therefore get the coefficients c_k for the best approximation directly:

$$c_k = \frac{(f, \varphi_k)}{(\varphi_k, \varphi_k)}, \qquad k = 0, 1, \ldots, N.$$

These c_k are called **Fourier coefficients**.

Using orthogonal polynomials φ_j has another advantage. Assume that the best approximation p_n^* of f among all polynomials of degree $\leq n$ has been computed. In order to determine the best approximation p_{n+1}^* of f among all polynomials of degree $\leq (n+1)$, we only need to know φ_{n+1} and then compute

$$c_{n+1}^* = \frac{(f, \varphi_{n+1})}{(\varphi_{n+1}, \varphi_{n+1})}.$$

We get

$$p_{n+1}^* = p_n^* + c_{n+1}^* \varphi_{n+1}.$$

We shall now show how to construct polynomials orthogonal with respect to a given scalar product. We start with an example.

Example 9.4.1   Construct polynomials P_0, P_1, P_2 of degree $0, 1$ and 2, respectively, which are orthogonal with respect to the scalar product

$$(f, g) = \int_{-1}^{1} f(x)g(x)\, dx.$$

We put

$$P_0 = 1, \quad P_1 = x + a_{11}, \quad P_2 = x^2 + a_{21}x + a_{22},$$

and determine the constants a_{ik} so that the polynomials become orthogonal.

The relation $(P_1, P_0) = 0$ gives

$$\int_{-1}^{1} (x + a_{11})\, dx = 0,$$

which leads to $a_{11} = 0$.

The orthogonality relations $(P_2, P_0) = 0$ and $(P_2, P_1) = 0$ give

$$\int_{-1}^{1} (x^2 + a_{21}x + a_{22})\, dx = 0,$$

$$\int_{-1}^{1} (x^2 + a_{21}x + a_{22})x\, dx = 0,$$

which have the solution $a_{22} = -1/3$, $a_{21} = 0$. Thus, $P_2 = x^2 - \frac{1}{3}$.

The construction of orthogonal polynomials can often be simplified by the following observation.

Lemma 9.4.2   Let P_i, $i = 0, 1, \ldots, N$, be polynomials such that degree $P_i = i$. The polynomials are an orthogonal system if and only if, for $k = 1, \ldots, N$, we have

$$(P_k, x^j) = 0, \qquad j = 0, 1, \ldots, k - 1.$$

Proof.   Assume that the polynomials $P_0, \ldots, P_N$ are an orthogonal system. Then they are linearly independent (Exercise 2). Therefore, x^j can be expressed as a linear combination of them:

$$x^j = \sum_{r=0}^{j} \alpha_{rj} P_r(x).$$

Hence, $(P_k, x^j) = \sum_{r=0}^{j} \alpha_{rj}(P_k, P_r) = 0$ for $j < k$.

The converse follows similarly, since an arbitrary polynomial P_i can be written as a linear combination of powers $1, x, \ldots, x^i$.   ∎

An alternative way of constructing orthogonal polynomials is to use the following theorem.

Theorem 9.4.3 To any scalar product, there exists an essentially unique sequence of orthogonal polynomials $\varphi_0, \varphi_1, \varphi_2, \ldots$, where φ_i is of degree i. The coefficients of the highest power can be chosen arbitrarily. When these coefficients are fixed, then the orthogonal system is uniquely determined. The polynomials satisfy a three–term recursion formula

$$\varphi_{n+1}(x) = (\alpha_n x - c_{nn})\varphi_n(x) - c_{n,n-1}\varphi_{n-1}(x), \quad n \geq 0,$$
$$\varphi_{-1}(x) = 0,$$
$$\varphi_0(x) = A_0,$$

where

$$c_{nn} = \frac{\alpha_n(x\varphi_n, \varphi_n)}{(\varphi_n, \varphi_n)}, \qquad c_{n,n-1} = \frac{\alpha_n(\varphi_n, \varphi_n)}{\alpha_{n-1}(\varphi_{n-1}, \varphi_{n-1})}.$$

In the discrete case, with the net $x_0, x_1, \ldots, x_M$, the last polynomial in the sequence is φ_M.

Proof. The polynomials are constructed according to the same principle as in Example 9.4.1.

Put
$$\varphi_0 = A_0,$$
$$\varphi_1 = A_1 x + b.$$

The requirement that $(\varphi_0, \varphi_1) = 0$ gives

$$(A_0, A_1 x + b) = A_0 A_1(1, x) + A_0 b(1, 1) = 0.$$

This defines b uniquely as soon as A_0 and A_1 have been chosen.

Assume that orthogonal polynomials $\varphi_0, \varphi_1, \ldots, \varphi_n$ have been constructed, and that $\varphi_j(x) = A_j x^j + \ldots$. Then $\varphi_{n+1} = A_{n+1} x^{n+1} + \ldots$ shall be determined so that

$$(\varphi_{n+1}, \varphi_j) = 0, \qquad j = 0, 1, \ldots, n.$$

Put $A_{n+1} = \alpha_n A_n$. Then $\varphi_{n+1} - \alpha_n x\varphi_n$ is a polynomial of degree $\leq n$, and

$$\varphi_{n+1}(x) = \alpha_n x\, \varphi_n(x) - \sum_{i=0}^{n} c_{ni}\varphi_i(x).$$

Now combine this with the orthogonality requirement above. For $j = 0, 1, \ldots, n$, we obtain

$$\alpha_n(x\varphi_n, \varphi_j) = \sum_{i=0}^{n} c_{ni}(\varphi_i, \varphi_j)$$

or, since $(\varphi_i, \varphi_j) = 0$ for $i \neq j$,

$$\alpha_n(x\varphi_n, \varphi_j) = c_{nj}(\varphi_j, \varphi_j).$$

But $(x\varphi_n, \varphi_j) = (\varphi_n, x\varphi_j) = 0$ for $j < n-1$, since $x\varphi_j$ is a polynomial of degree $< n$. Therefore, we get

$$c_{nj} = 0$$

for $j = 0, 1, \ldots, n - 2$, and

$$\varphi_{n+1} = \alpha_n x\varphi_n - c_{nn}\varphi_n - c_{n,n-1}\varphi_{n-1}, \quad n \geq 0,$$
$$\varphi_0 = A_0,$$
$$\varphi_{-1} = 0.$$

The requirement that $(\varphi_{n+1}, \varphi_n) = 0$ is satisfied precisely when

$$\alpha_n(x\varphi_n, \varphi_n) = c_{nn}(\varphi_n, \varphi_n).$$

This gives c_{nn}. Further, $(\varphi_{n+1}, \varphi_{n-1}) = 0$ precisely when

$$\alpha_n(x\varphi_n, \varphi_{n-1}) = c_{n,n-1}(\varphi_{n-1}, \varphi_{n-1}).$$

If we observe that

$$(\varphi_{n+1}, \varphi_{n+1}) = \alpha_n(\varphi_{n+1}, x\varphi_n)$$

for all $n \geq 0$ according to the recursion formula, we can simplify the expression for $c_{n,n-1}$. In particular, we have

$$(\varphi_n, x\varphi_{n-1}) = \frac{(\varphi_n, \varphi_n)}{\alpha_{n-1}} \quad \text{for } n \geq 1.$$

Using this, we get

$$c_{n,n-1} = \frac{\alpha_n(\varphi_n, \varphi_n)}{\alpha_{n-1}(\varphi_{n-1}, \varphi_{n-1})}.$$

Now, it only remains to show that the sequence ends with φ_M in the discrete case. The polynomial

$$P_{M+1}(x) = (x - x_0)(x - x_1) \ldots (x - x_M)$$

is orthogonal to $\varphi_0, \varphi_1, \ldots, \varphi_M$, since, for arbitrary $j \leq M$,

$$(P_{M+1}, \varphi_j) = \sum_{i=0}^{M} w_i P_{M+1}(x_i)\varphi_j(x_i) = 0.$$

The polynomial φ_{M+1} is essentially unique, and therefore P_{M+1} must be a multiple of φ_{M+1}, say $P_{M+1} = c\varphi_{M+1}$. Thus,

$$\|P_{M+1}\|_2 = \|c \cdot \varphi_{M+1}\|_2 = 0$$

and φ_{M+1} cannot be a member of the orthogonal system. ∎

Example 9.4.4   Construct polynomials $\varphi_0, \varphi_1, \varphi_2$, with leading coefficient 1, which are orthogonal with respect to the scalar product

$$(f, g) = \sum_{i=0}^{2} f(x_i)g(x_i),$$

where $x_0 = -\sqrt{3}/2, x_1 = 0$ and $x_2 = \sqrt{3}/2$. (The reason why we choose these points will be apparent in Section 9.6.)
    We use the formulas in Theorem 9.4.3, and put $A_0 = 1$ and $\alpha_n = 1$. We then have $\varphi_0 = 1$. This gives

$$(x\varphi_0, \varphi_0) = \sum_{i=0}^{2} x_i = 0.$$

Thus $c_{00} = 0$ and

$$\varphi_1 = x\varphi_0 = x.$$

For the next step, we compute

$$(x\varphi_1, \varphi_1) = \sum_{i=0}^{2} x_i^3 = 0,$$

$$(\varphi_0, \varphi_0) = 3,$$

$$(\varphi_1, \varphi_1) = \sum_{i=0}^{2} x_i^2 = 3/2.$$

This gives $c_{11} = 0$ and $c_{10} = 1/2$. Finally, we get

$$\varphi_2 = x^2 - 1/2.$$

An arbitrary finite interval $a \le x \le b$ can be transformed to $-1 \le t \le 1$ by the change of variables

$$x = \frac{b-a}{2}t + \frac{b+a}{2}.$$

Therefore, in the sequel, we assume that this has been done and we consider polynomials defined on $[-1, 1]$.

9.5 Legendre Polynomials

The Legendre polynomials P_n are defined as

$$P_n(x) = \frac{1}{2^n \cdot n!} \frac{d^n}{dx^n}(x^2 - 1)^n, \quad n \ge 1,$$
$$P_0(x) = 1.$$

Since $(x^2-1)^n$ is a polynomial of degree $2n$, it is seen that P_n is a polynomial of degree n. For instance, we get

$$P_0(x) = 1, \quad P_1(x) = x, \quad P_2(x) = \frac{1}{2}(3x^2 - 1),$$
$$P_3(x) = \frac{1}{2}(5x^3 - 3x), \quad P_4(x) = \frac{1}{8}(35x^4 - 30x^2 + 3).$$

The polynomials are illustrated in Figure 9.5.1.

In general, $|P_n(x)| \le 1$ holds for $x \in [-1, 1]$. Further, $P_n(x)$ has n distinct zeros in $(-1, 1)$. Between these zeros, P_n has maxima and minima, whose magnitude decrease towards the midpoint of the interval. It can be shown that the Legendre polynomials are orthogonal with the weight function $w(x) = 1$ on $[-1, 1]$, i.e.,

$$\int_{-1}^{1} P_i(x)P_k(x)\, dx = 0, \quad i \ne k.$$

Further,

$$\int_{-1}^{1} P_i^2(x)\, dx = \frac{2}{2i+1}.$$

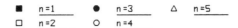

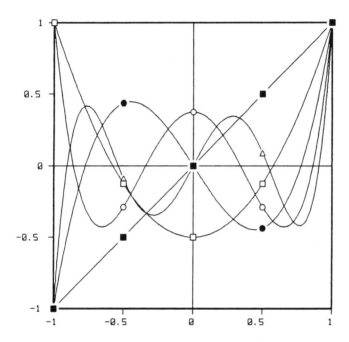

Figure 9.5.1   Legendre polynomials P_n.

Compare the Legendre polynomials P_0, P_1, P_2 with the polynomials that were constructed in Example 9.4.1. They are identical apart from a constant factor. The Legendre polynomials are normalized so that $P_n(1) = 1$ for all n.

These normalized polynomials satisfy the recursion

$$P_{n+1}(x) = \frac{2n+1}{n+1} x P_n(x) - \frac{n}{n+1} P_{n-1}(x), \quad n \geq 1,$$

$$P_0(x) = 1,$$

$$P_1(x) = x.$$

In particular, it follows that

$$P_n(-x) = (-1)^n P_n(x).$$

Example 9.5.2   Determine the second degree polynomial q that minimizes

$$\int_{-1}^{1} (x^3 - q(x))^2 \, dx.$$

In other words, we shall approximate x^3 by a lower degree polynomial using the least squares method. We put

$$q(x) = c_0 P_0(x) + c_1 P_1(x) + c_2 P_2(x),$$

where P_i are Legendre polynomials. The Fourier coefficients become

$$c_k = \frac{(x^3, P_k)}{(P_k, P_k)}, \qquad k = 0, 1, 2.$$

Here,

$$(x^3, P_0) = \int_{-1}^{1} x^3 \, dx = 0,$$

$$(x^3, P_1) = \int_{-1}^{1} x^4 \, dx = \frac{2}{5}, \quad (P_1, P_1) = \frac{2}{3}, \quad c_1 = \frac{3}{5},$$

$$(x^3, P_2) = \frac{1}{2} \int_{-1}^{1} x^3 (3x^2 - 1) \, dx = 0,$$

and we get

$$q(x) = \frac{3}{5}x.$$

We observe that $x^3 - q(x) = \frac{2}{5} P_3(x)$.

The following theorem holds in general.

Theorem 9.5.3   Given the scalar product

$$(f, g) = \int_{a}^{b} w(x) f(x) g(x) \, dx$$

and the corresponding orthogonal polynomials $p_0, p_1, \ldots, p_n$, all with leading coefficient equal to 1. Let q_n be an arbitrary nth degree polynomial with leading coefficient equal to 1. Then

$$\int_{a}^{b} w(x) q_n^2(x) \, dx$$

is minimized for $q_n = p_n$.

Proof. Express the nth degree polynomial q_n as a linear combination of the orthogonal polynomials $p_0, p_1, \ldots, p_n$:

$$q_n = p_n + \sum_{k=1}^{n-1} \alpha_k p_k.$$

Then

$$\int_a^b w(x) q_n^2(x)\, dx = \int_a^b w(x)\left(p_n(x) + \sum_{k=1}^{n-1} \alpha_k p_k(x)\right)^2 dx$$

$$= \int_a^b w(x) p_n^2(x)\, dx + \sum_{k=1}^{n-1} \alpha_k^2 \int_a^b w(x) p_k^2(x)\, dx.$$

This is minimized when $\alpha_1 = \alpha_2 = \ldots = \alpha_{n-1} = 0$.   ∎

9.6   Chebyshev Polynomials

The Chebyshev polynomials T_n are defined for $n \geq 0$ and $-1 \leq x \leq 1$ by

$$T_n(x) = \cos(n \arccos x).$$

In order to derive a three–term recursion for Chebyshev polynomials, we use the identity

$$e^{i\varphi} = \cos\varphi + i\sin\varphi.$$

Put $x = \cos\varphi$, $0 \leq \varphi \leq \pi$. Since $\cos\varphi = \text{Re}(e^{i\varphi})$, we get, for $n \geq 1$,

$$\begin{aligned}
T_{n+1}(x) + T_{n-1}(x) &= \cos((n+1)\varphi) + \cos((n-1)\varphi) \\
&= \text{Re}(e^{i(n+1)\varphi} + e^{i(n-1)\varphi}) = \text{Re}(e^{in\varphi}(e^{i\varphi} + e^{-i\varphi})) \\
&= \text{Re}(e^{in\varphi} \cdot 2\cos\varphi) = 2\cos\varphi\cos n\varphi \\
&= 2x T_n(x).
\end{aligned}$$

Therefore

$$\begin{aligned}
T_{n+1}(x) &= 2x T_n(x) - T_{n-1}(x), \quad n \geq 1, \\
T_0(x) &= 1, \\
T_1(x) &= x.
\end{aligned}$$

The formula also shows that the Chebyshev polynomials defined above are really polynomials. For instance, we get

$$\begin{aligned}
T_2(x) &= 2x^2 - 1, & T_3(x) &= 4x^3 - 3x, \\
T_4(x) &= 8x^4 - 8x^2 + 1, & T_5(x) &= 16x^5 - 20x^3 + 5x.
\end{aligned}$$

Using the recursion formula and induction, we see that the coefficient for x^n in $T_n(x)$ is 2^{n-1}, for $n \geq 1$. Further, since T_0 is an even function and T_1 is odd, we get

$$T_n(-x) = (-1)^n T_n(x).$$

The zeros of the Chebyshev polynomials x_k, $k = 0, 1, \ldots, n-1$, are called the **Chebyshev nodes** and are obtained from

$$T_n(x_k) = \cos(n \arccos x_k) = 0,$$

i.e.,

$$n \arccos x_k = (2k+1)\frac{\pi}{2},$$

or

$$x_k = \cos \frac{2k+1}{n} \cdot \frac{\pi}{2}.$$

The functions are illustrated in Figure 9.6.1.

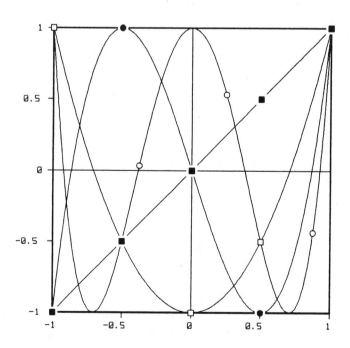

Figure 9.6.1   Chebyshev polynomials T_n.

From the figure, we also see that the zeros move closer to -1 and $+1$ for growing degree n. Between the zeros, T_n assumes the extremal values $+1$ and -1 at the points x'_k, $k = 0, 1, \ldots, n$:

$$T_n(x'_k) = \cos(n \arccos x'_k) = (-1)^k.$$

For the points x'_k, we thus have

$$n \arccos x'_k = k \cdot \pi,$$

or

$$x'_k = \cos \frac{k\pi}{n}.$$

In particular, $T_n(1) = 1$ for all n.

The Chebyshev polynomials are orthogonal with respect to the weight function $1/\sqrt{1 - x^2}$ in the interval $[-1, 1]$, since

$$(T_j, T_k) = \int_{-1}^{1} \frac{T_j(x) T_k(x)}{\sqrt{1 - x^2}}\, dx = \int_0^{\pi} T_j(\cos \varphi) T_k(\cos \varphi)\, d\varphi$$

$$= \int_0^{\pi} \cos j\varphi \cos k\varphi\, d\varphi$$

$$= \frac{1}{2} \int_0^{\pi} (\cos(j + k)\varphi + \cos(j - k)\varphi)d\varphi = \begin{cases} 0 & \text{for } j \neq k, \\ \frac{\pi}{2} & \text{for } j = k \neq 0, \\ \pi & \text{for } j = k = 0. \end{cases}$$

It can be shown that $T_0, T_1, \ldots, T_M$ are orthogonal with respect to the scalar product

$$(f, g) = \sum_{i=0}^{M} f(x_i) g(x_i),$$

where x_i are the zeros of $T_{M+1}(x)$. Note that, in the discrete case, the weight function is 1. In Example 9.4.4, we showed this for the case $M = 2$.

The following theorem expresses the important "minimax–property" of Chebyshev polynomials.

> **Theorem 9.6.2** Of all nth degree polynomials with 1 as the leading coefficient, $2^{-(n-1)}T_n$ has the smallest maximum norm on the interval $[-1, 1]$.

Proof. Assume the theorem is not true, i.e., that there exists a polynomial $p_n \neq 2^{-(n-1)}T_n$ such that

$$p_n(x) = x^n + \sum_{k=1}^{n} \alpha_k x^{n-k}$$

and

$$|p_n(x)| < 2^{-(n-1)} \quad \text{for all } x \in [-1,1].$$

$T_n(x)$ has extrema in the points $x_0' = 1, x_1', \ldots, x_{n-1}', x_n' = -1$. For the value of p_n in these points, we have:

$$p_n(x_0') < 2^{-(n-1)}T_n(x_0') = 2^{-(n-1)},$$

$$p_n(x_1') > 2^{-(n-1)}T_n(x_1') = -2^{-(n-1)},$$

$$p_n(x_2') < 2^{-(n-1)}T_n(x_2') = 2^{-(n-1)},$$

$$\text{etc.}$$

This can also be expressed as

$$p_n(x_0') - 2^{-(n-1)}T_n(x_0') < 0,$$

$$p_n(x_1') - 2^{-(n-1)}T_n(x_1') > 0,$$

$$p_n(x_2') - 2^{-(n-1)}T_n(x_2') < 0,$$

$$\text{etc.}$$

The polynomial $p_n - 2^{-(n-1)}T_n$, which has degree $\leq n-1$, therefore has a zero in each interval (x_k', x_{k+1}'), $k = 0, 1, \ldots, n-1$. Hence, in total, the polynomial has n zeros, which is impossible unless the polynomial is identically equal to zero. ∎

The following example shows how the minimax property of the Chebyshev polynomials can be used for approximation in the maximum norm.

Example 9.6.3 Determine the polynomial q, of degree ≤ 3, which minimizes

$$\max_{-1 \leq x \leq 1} |x^4 - q(x)|.$$

We use the fact that the Chebyshev polynomial $\frac{1}{8}T_4(x) = x^4 - x^2 + \frac{1}{8}$ is the fourth degree polynomial with leading coefficient 1 that has the smallest maximum norm on the interval $[-1, 1]$, and choose q so that

$$x^4 - q(x) = \frac{1}{8}T_4(x),$$

i.e.,

$$q(x) = x^2 - \frac{1}{8}.$$

Note that the error function oscillates between its extrema $\pm\frac{1}{8}$ in five points. We comment on this in connection with Theorem 9.7.6.

9.7   Maximum Norm—Chebyshev Approximation

In this section, we shall consider the continuous case when the norm is defined as

$$\|f\|_\infty = \max_{a \leq x \leq b} |f(x)|.$$

It can be shown that with respect to this norm there also exists for each $f \in C[a, b]$ a uniquely defined polynomial p_n^*, of degree $\leq n$, such that

$$E_n(f) = \|f - p_n^*\|_\infty \leq \|f - p_n\|_\infty$$

for all polynomials p_n of degree $\leq n$.

Further, according to a theorem of Weierstrass, any function $f \in C[a, b]$ can be approximated with arbitrarily good accuracy by polynomials, i.e., $E_n(f) \to 0$ as $n \to \infty$. There is also a known relation between the regularity properties of the function f and the behaviour of $E_n(f)$ for large n. For instance, we have $E_n(f) = O(1/n^k)$ as $n \to \infty$, for all f with k continuous derivatives.

It is considerably more difficult to determine p_n^* than to compute the best approximation in Euclidean norm. Therefore, we shall first describe a few simple methods for the *approximate* computation of p_n^*. We assume that the interval has been transformed to $[-1, 1]$.

One can approximate p_n^* by an interpolating polynomial. From the example in Section 5.2 (Figure 5.2.4), we see that *equidistant* interpolation points are a bad choice if we want the error to be small in the whole interval. In that example, there are large errors close to the endpoints of the interval. This suggests that we should put more interpolation points near the endpoints in order to force the interpolating polynomial to follow the given curve better there. Such a distribution of points is obtained by using the Chebyshev nodes. Interpolation in these nodes is called **Chebyshev interpolation**. To see that this is really a good choice of points, consider the error term in interpolation:

$$\frac{f^{(n+1)}(\xi(x))}{(n+1)!}(x - x_1)(x - x_2) \cdots (x - x_{n+1}).$$

To minimize the maximum norm of the error term of an arbitrary function f, which is $(n+1)$ times continuously differentiable in $[-1,1]$, we should choose the interpolation points $x_1, x_2, \ldots, x_{n+1}$ so that

$$\max_{-1 \le x \le 1} |(x - x_1)(x - x_2) \cdots (x - x_{n+1})|$$

is minimized. From Theorem 9.6.2, we see that this is achieved by taking

$$(x - x_1)(x - x_2) \cdots (x - x_{n+1}) = 2^{-n} T_{n+1}(x).$$

It can be shown that the magnitude of the error in Chebyshev interpolation with a polynomial of degree n is at most $4E_n(f)$ for $n \le 20$, and at most $5E_n(f)$ for $n \le 100$, if $f \in C[-1,1]$. Thus, the method gives a good approximation of the best maximum norm approximation to f. A practical way of computing the Chebyshev interpolation is shown in the following example.

Example 9.7.1   Compute approximations to the coefficients c_0 and c_1 which minimize

$$\phi(c_0, c_1) = \max_{0 \le x \le 1} |e^x - (c_0 + c_1 x)|.$$

We first make the transformation of variables

$$x = \frac{1}{2}(t+1),$$

to get the standard interval $[-1,1]$. This gives

$$\phi(c_0, c_1) = \max_{-1 \le t \le 1} |e^{(t+1)/2} - (c_0 + \frac{c_1}{2}(t+1))|.$$

Put $A_0 = c_0 + c_1/2$, $A_1 = c_1/2$ and $f(t) = e^{(t+1)/2}$. Now the problem is to compute A_0, A_1 so that

$$\max_{-1 \le t \le 1} |f(t) - (A_0 + A_1 t)|$$

is minimized.

Chebyshev interpolation means that A_0 and A_1 are determined from the requirement

$$A_0 + A_1 t_i = f(t_i), \qquad i = 0, 1,$$

where t_0, t_1 are the zeros of the Chebyshev polynomial $T_2(t) = 2t^2 - 1$, i.e., $t_0 = 1/\sqrt{2}$, $t_1 = -1/\sqrt{2}$.

In this example, where the degree of the approximating polynomial is low, it is easy to compute the coefficients A_0 and A_1 from the system of equations. On the other hand, if the degree of the approximating polynomial is high, then the computations become simpler if the interpolation problem is considered as a least squares problem and orthogonal polynomials are used. Therefore, let us use this technique here. We determine A_0, A_1 so that

$$\sum_{i=0}^{1} (f(t_i) - (A_0 + A_1 t_i))^2$$

is minimized.

The Chebyshev polynomials are orthogonal with respect to the scalar product

$$(g_1, g_2) = \sum_{i=0}^{1} g_1(t_i) g_2(t_i).$$

Therefore, express the approximating polynomial in terms of Chebyshev polynomials:

$$A_0 + A_1 t = A_0 T_0 + A_1 T_1.$$

The coefficients A_0, A_1 are now obtained directly as Fourier coefficients:

$$A_k = \frac{(f, T_k)}{(T_k, T_k)} = \frac{\sum_{i=0}^{1} f(t_i) T_k(t_i)}{\sum_{i=0}^{1} T_k^2(t_i)}, \qquad k = 0, 1.$$

Thus,

$$A_0 = \frac{1}{2} \sum_{i=0}^{1} f(t_i) = \frac{1}{2} \left(e^{(\sqrt{2}+1)/(2\sqrt{2})} + e^{(\sqrt{2}-1)/(2\sqrt{2})} \right),$$

$$A_1 = \sum_{i=0}^{1} f(t_i) t_i = \frac{1}{\sqrt{2}} \left(e^{(\sqrt{2}+1)/(2\sqrt{2})} - e^{(\sqrt{2}-1)/(2\sqrt{2})} \right),$$

and

$$c_0 = A_0 - A_1 \approx 0.9112,$$
$$c_1 = 2A_1 \approx 1.6833.$$

The corresponding error function has its maximum value equal to 0.124 at $x = 1$, while the best maximum norm approximation by a straight line has a maximal error 0.106.

An alternative way of approximating p_n^* is to use approximation with the Euclidean norm and the weight function $1/\sqrt{1-x^2}$ on $[-1, 1]$. Intuitively, we see that this weight function must have the effect of making the error in the approximation small close to -1 and $+1$. Let us study this technique in more detail. We first define the notion of an orthogonal expansion.

Definition 9.7.2 Let $\varphi_0, \varphi_1, \varphi_2, \ldots$ be an orthogonal system of polynomials with respect to the scalar product

$$(g_1, g_2) = \int_a^b w(x) g_1(x) g_2(x)\, dx.$$

The expansion $\sum_{i=0}^{\infty} c_i \varphi_i(x)$, where $c_i = (f, \varphi_i)/(\varphi_i, \varphi_i)$, is called an **orthogonal expansion of the function** f.

From the discussion earlier in this chapter, we know that the partial sums $p_n(x) = \sum_{i=0}^{n} c_i \varphi_i(x)$ of the orthogonal expansion are the best approximation to f among all polynomials of degree $\leq n$, with respect to the correspondingly weighted Euclidean norm $\|\cdot\|_{2,w}$.

Now let $p_n^*(x)$, as before, be the best approximation to f in *maximum norm* among all polynomials of degree at most n. Then

$$\|f - p_n\|_{2,w}^2 \leq \|f - p_n^*\|_{2,w}^2 = \int_a^b w(x)(f(x) - p_n^*(x))^2\, dx$$

$$\leq \max_{a \leq x \leq b} |f(x) - p_n^*(x)|^2 \int_a^b w(x)\, dx$$

$$= E_n^2(f) \int_a^b w(x)\, dx.$$

Since, according to Weierstrass, $E_n(f) \to 0$ as $n \to \infty$, we also get

$$\|f - p_n\|_{2,w} \to 0 \qquad \text{as } n \to \infty.$$

Only in certain special cases does p_n converge uniformly to f as $n \to \infty$. This is the case, e.g., if f is twice continuously differentiable in $[-1, 1]$ and the φ_i are chosen as Chebyshev polynomials. A truncated orthogonal expansion $\sum_{i=0}^{n} c_i T_i$, where the T_i are Chebyshev polynomials, should therefore give a good approximation to p_n^*. It has been shown that

$$\left\| f - \sum_{i=0}^{n} c_i T_i \right\|_\infty < \left(4 + \frac{4}{\pi^2} \log n\right) E_n(f)$$

for $f \in C[-1, 1]$. For values up to $n = 1000$, the constant in front of $E_n(f)$ is at most 5.2. Thus, the loss of accuracy in the approximation to p_n^* by $\sum_{i=0}^n c_i T_i$ is acceptable.

Example 9.7.3 For the problem in Example 9.7.1, we get, using this technique,

$$A_0 = \frac{(e^{(t+1)/2}, T_0)}{(T_0, T_0)} = \frac{1}{\pi} \int_{-1}^1 \frac{e^{(t+1)/2}}{\sqrt{1 - t^2}} \, dt,$$

$$A_1 = \frac{(e^{(t+1)/2}, T_1)}{(T_1, T_1)} = \frac{2}{\pi} \int_{-1}^1 \frac{t e^{(t+1)/2}}{\sqrt{1 - t^2}} \, dt.$$

After the substitution of variables $t = \cos \varphi$, and using a standard program for the integration, we get

$$A_0 \approx 1.7534,$$
$$A_1 \approx 0.8504,$$

and

$$c_0 = A_0 - A_1 \approx 0.9030,$$
$$c_1 = 2A_1 \approx 1.7008.$$

As a matter of fact, it is quite easy to determine *exactly* the best maximum norm approximation to a given monotone function by a *straight line*. The line shall be chosen so that the error function assumes extremal values with alternating signs in three points, namely, the endpoints of the interval and a point in the interior (Figure 9.7.5). It is obvious that the maximal error cannot be made smaller by moving this line.

Example 9.7.4 Compute c_0, c_1 so that

$$\max_{0 \le x \le 1} \left| e^x - (c_0 + c_1 x) \right|$$

is minimized.

We use the fact that the error curve shall assume the extremal value d with alternating signs in the points $0, \xi, 1$.

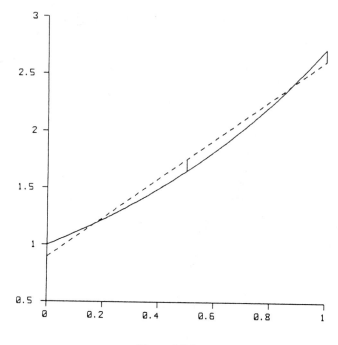

Figure 9.7.5

With $r(x) = e^x - (c_0 + c_1 x)$ we then require

$$r(0) = d,$$
$$r(\xi) = -d,$$
$$r(1) = d,$$

and

$$r'(\xi) = 0.$$

This gives four equations for the four unknowns c_0, c_1, d and ξ:

$$1 - c_0 = d,$$
$$e^\xi - c_0 - c_1 \xi = -d,$$
$$e - c_0 - c_1 = d,$$
$$e^\xi - c_1 = 0.$$

After simple calculations, we get

$$c_0 = \frac{1}{2}(e - (e-1)\log(e-1)) \approx 0.8941,$$

$$c_1 = e - 1 \approx 1.7183,$$

$$\xi = \log(e-1) \approx 0.54,$$

$$d = 1 - c_0 \approx 0.1059.$$

In Example 9.6.3, we saw that the error function in the approximation by a polynomial of degree $n = 3$ oscillated between extremal values in $(n+2) = 5$ points.

The following holds in general.

Theorem 9.7.6   Assume that $f \in C[a,b]$. Then p_n^* is the best maximum norm approximation to f among all polynomials of degree $\leq n$ if and only if there are points $x_0, x_1, \ldots, x_{n+1}$ such that

$$|f(x_k) - p_n^*(x_k)| = \|f - p_n^*\|_\infty, \qquad k = 0, 1, \ldots, n+1,$$

and

$$f(x_k) - p_n^*(x_k) = -(f(x_{k+1}) - p_n^*(x_{k+1})), \qquad k = 0, 1, \ldots, n.$$

Thus, the error function $f - p_n^*$ alternates between $\pm\|f - p_n^*\|$ in at least $(n+2)$ points. This so-called **alternation property** is used in algorithms for constructing p_n^*. E.g., we did so for $n = 1$ in Example 9.7.4. For $n > 1$, it is not as easy. Assume, for instance, that e^x shall be approximated by a second degree polynomial $p_2(x) = c_0 + c_1 x + c_2 x^2$. Denote by x_0, x_1, x_2, x_3 the extrema of the error function $e^x - p_2^*$ corresponding to the best approximation. It is easy to show that $x_0 = 0$ and $x_3 = 1$. The alternation property gives

$$1 - c_0 = d,$$
$$e^{x_1} - (c_0 + c_1 x_1 + c_2 x_1^2) = -d,$$
$$e^{x_2} - (c_0 + c_1 x_2 + c_2 x_2^2) = d,$$
$$e - (c_0 + c_1 + c_2) = -d.$$

If x_1 and x_2 were known, then we would have a linear system of equations for determining c_0, c_1, c_2 and d. This can be utilized by determining the alternation points iteratively as follows (the **Remes algorithm**):

1. "Guess" $x_1^{(0)}$ and $x_2^{(0)}$.
2. Compute c_0, c_1, c_2 and d from the system of equations above.
3. Compute interior extrema $x_1^{(1)}$ and $x_2^{(1)}$ for the corresponding error function.
4. If the error function alternates according to Theorem 9.7.6, stop. Otherwise, replace $x_i^{(0)}$ by $x_i^{(1)}$, $i = 1, 2$, and repeat from 2.

In the guessing of starting values, the approximate methods described earlier in this section can be used, e.g., Chebyshev interpolation.

Exercises

1. Let p_1 and p_2 be the best least squares approximations to f_1 and f_2, respectively, among polynomials of degree n. Show that $\alpha_1 p_1 + \alpha_2 p_2$ is the best least squares approximation to the function $\alpha_1 f_1 + \alpha_2 f_2$.

2. Let $\varphi_0, \varphi_1, \ldots, \varphi_N$ be an orthogonal system. Prove the following generalization of the Pythagorean law:

$$\| \sum_{j=0}^{N} c_j \varphi_j \|_2^2 = \sum_{j=0}^{N} c_j^2 \|\varphi_j\|_2^2.$$

Use this to show that the functions in an orthogonal system are linearly independent.

3. Let $f^* = \sum_{j=0}^{N} c_j^* \varphi_j$ be the best least squares approximation to a given function f. Use the Pythagorean law to show that

$$\|f - f^*\|_2^2 = \|f\|_2^2 - \|f^*\|_2^2.$$

4. For the computation of $f(x) = \sum_{k=0}^{n} a_k P_k(x)$, where the $P_k(x)$ are Legendre polynomials, the following recursion formula can be used:

$$b_{n+2} = b_{n+1} = 0,$$
$$b_k = a_k + \frac{2k+1}{k+1} x b_{k+1} - \frac{k+1}{k+2} b_{k+2}, \quad k = n, n-1, \ldots, 0.$$

Show that $f(x) = b_0$.

5. For the computation of $f(x) = \sum_{k=0}^{n} a_k T_k(x)$, where the $T_k(x)$ are Chebyshev polynomials, the following recursion formula can be used:

$$b_{n+2} = b_{n+1} = 0$$
$$b_k = a_k + 2xb_{k+1} - b_{k+2}, \qquad k = n, n-1, \ldots, 0.$$

Show that $f(x) = b_0 - xb_1$.

6. Let $x = \cos \varphi$, and define

$$U_n(x) = \frac{\sin(n+1)\varphi}{\sin \varphi}, \qquad n = 0, 1 \ldots.$$

a) $\{U_n(x)\}_0^\infty$ satisfy a recursion formula of the type

$$U_{n+1}(x) = (a_{n+1}x - b_{n+1})U_n(x) - c_{n+1}U_{n-1}(x).$$

   Show this, and determine the coefficients.
b) Show that $U_n(x)$ is a polynomial in x of degree n.
c) The functions $U_n(x)$, $n = 0, 1, \ldots$, form an orthogonal system on the interval $[-1, 1]$ with respect to a certain weight function $w(x)$. Determine $w(x)$. ($U_n(x)$ are called **Chebyshev polynomials of the second kind.**)

7. For the function $f(x)$, the following holds: $|f^{(n)}(x)| \leq M$ for all $x \in [-1, 1]$. Show that there is a polynomial P_{n-1}, of degree $n - 1$, such that

$$|f(x) - P_{n-1}(x)| \leq \frac{M}{2^{n-1} \cdot n!} \qquad \text{for all } x \in [-1, 1].$$

8. Use the minimax property of the Chebyshev polynomials to show that, for each polynomial $p_n(x) = \sum_{k=0}^{n} c_k x^k, c_n \neq 0$, there is a point $-1 \leq \xi \leq 1$ such that

$$|p_n(\xi)| \geq |c_n| \cdot 2^{-(n-1)}.$$

9. Approximate the function $f(x) = \sqrt[3]{x}$ by a straight line on the interval $[0, 1]$:
   a) in the least squares sense with the weight function 1;
   b) in maximum norm. In both cases, give the norm of the error function for the best approximation.

10. Approximate the function $f(x) = (2x - 1)^4$ by a second degree polynomial on the interval $[0, 1]$:
   a) in the least squares sense with the weight function 1;
   b) in maximum norm. In both cases, give the norm of the error function for the best approximation.

11. a) Compute orthogonal polynomials $\varphi_k(x)$, $k = 0, 1, 2$, with leading coefficient 1, for the weight function $w(x) = x^2$ in the interval $-1 \le x \le 1$.
   b) Compute the second degree polynomial $p_2(x)$ which minimizes

$$\int_{-1}^{1} x^2 (e^x - p_2(x))^2 \, dx.$$

   c) Use the results from Exercises 2 and 3 to compute

$$\|e^x - p_2(x)\|_2^2 = \int_{-1}^{1} x^2 (e^x - p_2(x))^2 \, dx.$$

12. a) Determine polynomials $\varphi_k(x)$, $k = 0, 1, 2, 3, 4$, orthogonal on the net $\{-2, -1, 0, 1, 2\}$, such that the degree of φ_k is equal to k and the coefficient of x^k is 1. Use the recursion $\varphi_k = (x + b_k)\varphi_{k-1} + c_k\varphi_{k-2}$. First, show that $b_k = 0$.
   b) Is it possible to construct a polynomial φ_5, of degree 5, orthogonal to $\varphi_0, \dots, \varphi_4$?
   c) Use the polynomials from a) to determine the polynomial $p^*(x)$ of degree 4 that minimizes

$$\sum_{k=1}^{5} (f(x_k) - p(x))^2,$$

   where f is given in the table

x	-2	-1	0	1	2
$f(x)$	29	7	5	5	13

References

The presentation of the least squares method is inspired by the corresponding sections in

G. Dahlquist and Å. Björck, *Numerical Methods*, Prentice–Hall, Engle-wood Cliffs, New Jersey, 1974.

The first book that gave useful approximations for the computation of elementary functions on computers was

C. Hastings, *Approximations for Digital Computers*, Princeton University Press, Princeton, New Jersey, 1955.

The choice of approximating functions was made using intuition and great artistic skill. Now there are systematic methods for the computation of best approximations with both polynomials and rational functions. Good approximations of these kinds are given in

J. F. Hart et al., *Computer Approximations*, John Wiley, New York, 1968.

An easy introduction to approximation in different norms is given in

T. J. Rivlin, *An Introduction to the Approximation of Functions*, Dover Publications, Inc., New York, 1981.

and a more extensive treatment in

G. A. Watson, *Approximation Theory and Numerical Methods*, John Wiley and Sons, Chichester, 1980.

For those who, in addition, want an extensive description of interpolation and approximation with spline functions, we recommend

M. J. D. Powell, *Approximation Theory and Methods*, Cambridge University Press, Cambridge, 1981.

10   Differential Equations

10.1   Introduction

A **differential equation** is an equation involving an unknown function and one or several of its derivatives. Let $f(x, y)$ denote a given real valued function of two variables. In this chapter, we will consider **ordinary differential equations of first order**, which can be written

$$y' = f(x, y),$$

where $y(x)$ is the unknown function that we want to determine. In general, such an equation has infinitely many solutions.

Example 10.1.1   Let the given function f be $f(x, y) = y$. The differential equation

$$y' = y$$

is satisfied by $y(x) = Ce^x$ for all values of C, see Figure 10.1.2.

In order to decide which of the infinitely many solutions that one is really looking for, one must give an extra condition of the type $y(a) = \alpha$. This is called an **initial condition**, and the problem of solving

$$y' = f(x, y), \qquad y(a) = \alpha,$$

is called an **initial value problem**.

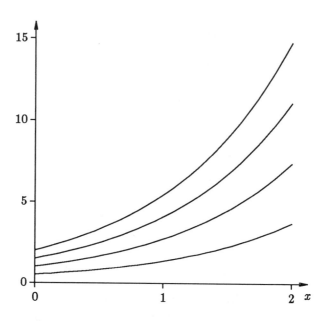

Figure 10.1.2   Solutions of $y' = y$.

Example 10.1.3   The initial value problem

$$y' = y, \qquad y(0) = 1,$$

has the solution $y = e^x$.

Initial value problems for differential equations are often models of dynamic processes in applied sciences. We give a simple example.

Example 10.1.4   Consider a box resting on a firm foundation, and which is connected to a wall by a spring.

Introduce a coordinate axis in such a way that the box is at the origin when the spring exerts no force. If we remove the spring from the origin and release it, then the spring will pull the box back towards the origin. Let $y(t)$ denote the position of the box at time t. Hooke's law states that the spring exerts a force on the box that is proportional to y. The friction between the box and the foundation is proportional to the speed $y'(t)$. Newton's second law of motion states that the mass times the acceleration is equal to the force. We then have

$$my'' = -qy - \phi y',$$

where m is the mass of the box, and q and ϕ are proportionality constants.

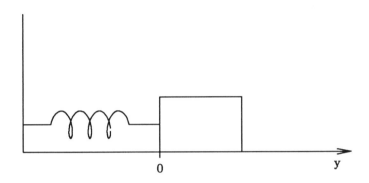

Figure 10.1.5

We immediately see that, in order to solve the differential equation, we must know the initial position and the initial velocity of the box. Thus the initial conditions are

$$y(0) = \alpha, \quad y'(0) = \beta,$$

for some given constants α and β.

The equation in the preceding example is an **ordinary differential equation of second order**, since it has a second derivative. To solve such an equation, one must have two additional conditions. By introducing new unknown functions any higher order differential equation can be written as a **system of first order differential equations**.

Example 10.1.6   In Example 10.1.4, put $y_1 = y$, and $y_2 = y'$. Then we can write the differential equation

$$my_2' = -qy_1 - \phi y_2.$$

We get another equation from the observation that $y_1' = y' = y_2$. The system becomes

$$\begin{cases} y_1' = y_2, \\ my_2' = -qy_1 - \phi y_2, \end{cases} \qquad \begin{cases} y_1(0) = \alpha, \\ y_2(0) = \beta, \end{cases}$$

or, with matrix notation,

$$\begin{pmatrix} y_1' \\ y_2' \end{pmatrix} = \begin{pmatrix} 0 & 1 \\ -q/m & -\phi/m \end{pmatrix} \begin{pmatrix} y_1 \\ y_2 \end{pmatrix}, \qquad \begin{pmatrix} y_1(0) \\ y_2(0) \end{pmatrix} = \begin{pmatrix} \alpha \\ \beta \end{pmatrix}.$$

This system is of the form $y' = f(t, y)$, where now y and f are vector valued functions

$$y = \begin{pmatrix} y_1 \\ y_2 \end{pmatrix}, \qquad f(t, y) = \begin{pmatrix} f_1(t, y_1, y_2) \\ f_2(t, y_1, y_2) \end{pmatrix}.$$

In the sequel, we write $y' = f(x, y)$, and then we presuppose that the equation is scalar, i.e., that y is a function from R to R. Note, however, that most of the material in this chapter can, after small modifications, be applied to systems of differential equations.

For completeness we state here a theorem that shows when an initial value problem has a unique solution.

Theorem 10.1.7   Let the function $f(x, y)$ be defined and continuous for all points (x, y) such that $a \le x \le b$, $-\infty < y < \infty$, where a and b are finite. If f satisfies a **Lipschitz condition**, i.e., if there exists a constant L such that

$$|f(x, y) - f(x, y^*)| \le L\,|y - y^*|$$

for $a \le x \le b$ and all y, y^*, then for any initial value α there exists a unique solution of the initial value problem

$$y' = f(x, y), \qquad y(a) = \alpha.$$

L is called the **Lipschitz constant**.

The statement of the theorem can also be formulated as follows: Through all points (x_0, y_0), where x_0 is in the interval $[a, b]$, there passes exactly one curve that is a solution of the equation $y' = f(x, y)$.

Example 10.1.8   Consider the initial value problem

$$y' = f(x, y) = xy, \qquad y(0) = 1,$$

and assume that we want to solve the equation in the interval $[0, 1]$. We immediately get

$$|xy - xy^*| \leq |y - y^*|.$$

The Lipschitz constant is $L = 1$. More generally, it is easy to show that if

$$\left| \frac{\partial f}{\partial y} \right| \leq L,$$

in the interval considered, then L can be used as the Lipschitz constant.

The problem from structural mechanics that was described briefly in Chapter 1 led to a **boundary value problem for an ordinary differential equation**. Later in this chapter, we will study numerical methods for solving problems of the type

$$y'' = f(x, y, y'), \qquad y(a) = \alpha, \quad y(b) = \beta,$$

where $[a, b]$ is an interval.

10.2 Numerical Solution of Initial Value Problems

Most differential equations that arise in applications cannot be solved using analytical methods. Therefore, it is necessary to make numerical approximations. The numerical methods we will describe are based on the following idea: Since we cannot determine the function $y(x)$ for all x in an interval $[a, b]$, we will have to be satisfied with computing *approximations* y_i of $y(x_i)$ for some points $(x_i)_{i=0}^{N}$ in the interval. We assume that the points are equidistant, i.e., that

$$x_i = a + ih, \qquad i = 0, \ldots, N,$$

where the **step length** h is defined as

$$h = \frac{(b - a)}{N},$$

for some integer N.

Figure 10.2.1   The interval $[a, b]$ divided into N subintervals.

The initial value gives us $y_0 = y(x_0) = \alpha$ at the point $x = x_0 = a$. Now we want to compute y_1, which is an approximation of $y(x_1)$. We will do this by **discretizing** the differential equation. To make the derivation somewhat more general, we assume that we know y_n (the approximation of $y(x_n)$), and that we want to compute y_{n+1}. The differential equation at the point $x = x_n$ is

$$y'(x_n) = f(x_n, y(x_n)).$$

We first replace the derivative by a difference quotient

$$y'(x_n) \approx \frac{y(x_{n+1}) - y(x_n)}{h},$$

and the differential equation becomes

$$\frac{y(x_{n+1}) - y(x_n)}{h} \approx f(x_n, y(x_n)).$$

Then, we replace $y(x_n)$ and $y(x_{n+1})$ by y_n and y_{n+1} in the difference quotient and the right hand side, and we get

$$y_{n+1} = y_n + hf(x_n, y_n).$$

This is a classical method for the numerical solution of initial value problems, and it is called Euler's method.

Euler's method:

$$y_{n+1} = y_n + hf(x_n, y_n), \qquad y_0 = \alpha.$$

With this **recursion formula**, we can successively compute $y_1, y_2, \ldots$, which are approximations of $y(x_1), y(x_2), \ldots$.

We want to emphasize that Euler's method is hardly ever used in practice, as there are more accurate and more efficient (but, at the same time, more complicated) methods. Euler's method is simple, and that is why we use it for introducing the basic concepts in the numerical solution of initial value problems.

Example 10.2.2   In the sequel, we will illustrate several methods by numerically solving the initial value problem

$$y' = f(x, y) = xy, \qquad y(0) = 1.$$

This problem has the analytical solution $y(x) = e^{x^2/2}$, and we will use this to check the accuracy of the numerical solution.

Assume that we want to compute an approximation of $y(0.4)$. Euler's method applied to this equation is

$$y_{n+1} = y_n + hx_n y_n,$$

and the initial value gives $y_0 = 1$. We first let $h = 0.2$. Then, we have $x_1 = 0.2$ and $x_2 = 0.4$, and we see that it is y_2 that is the approximation of $y(0.4)$ we are looking for. We now use Euler's method:

$$y_1 = y_0 + hx_0 y_0 = 1 + 0.2 \cdot 0 \cdot 1 = 1,$$
$$y_2 = y_1 + hx_1 y_1 = 1 + 0.2 \cdot 0.2 \cdot 1 = 1.04.$$

We summarize the computations in a table, where we also give the analytical solution:

n	x_n	y_n	$y(x_n)$	Error
0	0	1	1	0
1	0.2	1	1.020	0.020
2	0.4	1.040	1.083	0.043

Next, we halve the step length to $h = 0.1$, and compute a new approximation of $y(0.4)$. Note that it is now y_4 that we want to compute. The result is

n	x_n	y_n	$y(x_n)$	Error
0	0	1	1	0
1	0.1	1.000	1.005	0.005
2	0.2	1.010	1.020	0.010
3	0.3	1.030	1.046	0.016
4	0.4	1.061	1.083	0.022

If we compare the two tables, we see that the error was halved when the step length was halved. This indicates that the truncation error is proportional to the step length h.

Euler's method has a geometric interpretation: We start at the point (x_0, y_0), and approximate the solution curve by the tangent at the point (x_0, y_0). We compute the slope $y'(x_0)$ of the tangent directly from the differential equation. We continue along this tangent until we reach $x = x_1 = x_0 + h$. The corresponding y-value is y_1. Through the point (x_1, y_1), there is a solution curve (which, however, does not correspond to the given initial value). Similarly, we approximate this curve by a tangent through the point (x_1, y_1) and continue along this tangent until we reach $x = x_2$, and so on.

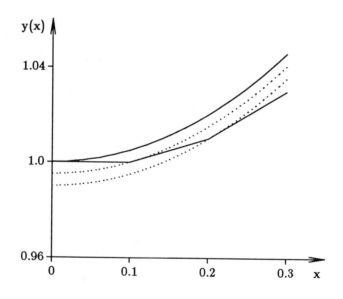

Figure 10.2.3   Euler's method corresponds to a polygonal curve.

10.3   Local and Global Truncation Error

The main sources of error in the numerical solution of differential equations are truncation error and rounding error. We distinguish between local and global truncation error.

The **local truncation error** at the point x_{n+1} is the difference between the computed value y_{n+1} and the value at the point x_{n+1} on
the solution curve that goes through the point (x_n, y_n).

From Figure 10.2.3, we see that the local truncation error in Euler's
method is the deviation after each step between a solution curve and its
tangent.

Example 10.3.1   We will show that if the solution of the differential
equation is twice continuously differentiable, then the local truncation
error in Euler's method is $O(h^2)$. According to the definition, we shall
start from the solution of the initial value problem

$$y' = f(x, y), \qquad y(x_n) = y_n.$$

The Taylor expansion around $x = x_n$ is

$$y(x_n + h) = y(x_n) + hy'(x_n) + \frac{h^2}{2}y''(\xi) = y_n + hf(x_n, y_n) + \frac{h^2}{2}y''(\xi),$$

where $\xi \in [x_n, x_n + h]$. The local truncation error is

$$y_{n+1} - y(x_n + h) = -\frac{h^2}{2}y''(\xi) = O(h^2).$$

The **global truncation error** in the point x_{n+1} is defined as

$$y_{n+1} - y(x_{n+1}),$$

where $y(x)$ denotes the solution of the given initial value problem.

In Figure 10.2.3, the global truncation error is equal to the distance from
a point on the solid curve to the corresponding point on the polygonal curve.
In general, if the local truncation error of a numerical method is $O(h^{p+1})$,
then the global error is $O(h^p)$.

Example 10.3.2   The global truncation error in Euler's method is $O(h)$. This was shown experimentally in Example 10.2.2, and we will now show it theoretically. We first put $\epsilon_n = y_n - y(x_n)$. By a Taylor expansion around $x = x_n$, we have

$$y(x_n + h) = y(x_n) + hy'(x_n) + \frac{h^2}{2}y''(\xi)$$

$$= y(x_n) + hf(x_n, y(x_n)) + \frac{h^2}{2}y''(\xi).$$

We subtract this equation from $y_{n+1} = y_n + hf(x_n, y_n)$, and get

$$\epsilon_{n+1} = \epsilon_n + h(f(x_n, y_n) - f(x_n, y(x_n))) - \frac{h^2}{2}y''(\xi).$$

Now, if we assume that f satisfies a Lipschitz condition with constant L, and that $|y''(x)| \le M$ for all x in the interval of interest (i.e., the interval where we want to determine the solution of the initial value problem), then we can make the estimate

$$|\epsilon_{n+1}| \le (1 + hL)\,|\epsilon_n| + \frac{h^2}{2}M.$$

Put $A = (1 + hL)$ and $B = Mh^2/2$. A simple induction proof gives

$$|\epsilon_n| \le A^n\,|\epsilon_0| + \left(\sum_{k=0}^{n-1} A^k\right)B,$$

i.e., for $A \ne 1$,

$$|\epsilon_n| \le \frac{A^n - 1}{A - 1}B,$$

since $\epsilon_0 = 0$. If we use the inequality $1 + x \le e^x$, we now get

$$A^n = (1 + hL)^n \le e^{Lnh} = e^{L(x_n - x_0)}.$$

Inserting this in the inequality for ϵ, we finally get

$$|\epsilon_n| \le \frac{hM}{2L}(e^{L(x_n - x_0)} - 1),$$

which shows that the global truncation error is $O(h)$.

The estimate of the global truncation error that was derived in Example 10.3.2 cannot be used for practical error estimation, since in most cases it is much too pessimistic. Also, in practice, one does not know M or L. The estimate has a certain theoretical interest, however. For instance, we can see that

$$\lim_{h \to 0} \epsilon_n = 0,$$

if we keep $x_n = x_0 + nh$ fixed as h tends to zero. This means that Euler's method is convergent in the sense that the error in the approximate solution for a fixed x-value tends to zero as the step length tends to zero. It can also be shown that, under certain assumptions on f, there is an expansion in powers of h of the global truncation error.

The global truncation error in Euler's method can be written

$$y_n - y(x_n) = a_1 h + a_2 h^2 + a_3 h^3 + \dots,$$

for some constants $a_1, a_2, \dots$ which do not depend on h.

This means that we can use Richardson extrapolation to obtain better approximations of the solution. At the same time, we can get approximations of the truncation error of the solution (as in the case of numerical integration). We first solve the initial value problem numerically with step length h over an interval $[a, b]$, and get a table of approximations of the solution at the points $x_n = x_0 + nh$, $n = 0, 1, \dots, N$. Then the same problem is solved with step length $h/2$ and we get a table with twice as many entries.

Now, there are two approximations of the same function value in $N = (b - a)/h$ points. These approximations can be refined using Richardson extrapolation. Since we have halved the step length, the correction term in the extrapolation is

$$\frac{\Delta}{1}$$

(see Chapter 6). If we continue to halve the step length, we can perform repeated Richardson extrapolation. The correction terms are

$$\frac{\Delta}{1}, \quad \frac{\Delta}{3}, \quad \frac{\Delta}{7}, \dots\dots$$

Example 10.3.3   We saw in Example 10.2.2 that the global error in
Euler's method applied to the equation $y' = xy$ was halved when the
step length was halved. This we have also shown theoretically. We will
now refine the approximations using Richardson extrapolation. We
introduce the notation $y(x; h)$, which is to be read "the approximation
of $y(x)$ obtained using a certain numerical method (in this example,
Euler's method) applied with the step length h." The results from
Example 10.2.2 are summarized in the following table, where we also
give the extrapolated values.

x	$y(x; 0.2)$	$y(x; 0.1)$	y_{extr}	Error
0	1	1	1	0
0.1		1.0000		
0.2	1	1.0100	1.0200	0.0002
0.3		1.0302		
0.4	1.0400	1.0611	1.0822	0.0011

The extrapolated entries are computed from the formulas for Richardson extrapolation, which can be written

$$y_{\text{extr}} = y(x; h/2) + (y(x; h/2) - y(x; h)).$$

Note that the errors in the extrapolated values are of the order
of magnitude 10^{-4}, whereas the errors without extrapolation are approximately 10^{-2}.

This procedure is called **passive Richardson extrapolation**, where
"passive" refers to the fact that the extrapolation is performed only after we
have computed approximations of $y(x)$ over the whole interval of interest. It
is also possible to use **active** extrapolation, i.e., compute approximations of
$y(x)$ with two different step lengths, extrapolate, and use the extrapolated
value for the next step in the numerical procedure. This method suffers
from stability problems, however, and therefore it should not be used.

Example 10.3.4   We again use the equation $y' = xy$, $y(0) = 1$. When we have computed approximations of $y(0.4)$ with Euler's
method and the step lengths $h = 0.2, 0.1$ and 0.05, we can use re-
peated Richardson extrapolation. The following table is obtained.

h	$y(0.4; h)$	$\Delta/1$	$\Delta/3$	
0.2	1.040000			
		0.021106		
0.1	1.061106		1.082212	
		0.010937		0.000256
0.05	1.072043		1.082980	1.083236

The truncation error in the rightmost entry of the table can be estimated as

$$|R_T| \lesssim |1.082212 - 1.082980| = 0.000768.$$

The analytical solution gives $y(0.4) \approx 1.083287$, and the true truncation error is approximately $5.1 \cdot 10^{-5}$.

10.4 Runge–Kutta Methods

Since the global truncation error in Euler's method is $O(h)$, it is often necessary to use a very short step length to get the desired accuracy in the approximate solution. There are several ways to derive more accurate methods. For instance, if the derivative in the differential equation is replaced by a central difference, instead of a forward difference then we have the **midpoint method** with global error $O(h^2)$. Unfortunately, this method is *unstable*. We analyze the midpoint method in Section 10.6.

We will now derive a method starting from the geometrical interpretation of Euler's method. In Figure 10.2.3, we see that the error in Euler's method is large due to the fact that we go along the direction of the tangent at the point (x_n, y_n) for a whole step, while the solution curve starts to deviate from this direction by a considerable amount during the step. If we take the average of the tangent directions at the points (x_n, y_n) and $(x_{n+1}, y_{n+1}^{(e)})$, where $y_{n+1}^{(e)}$ denotes the result for $x = x_{n+1}$ with Euler's method, we should be able to make a correction for the bending of the curve; see Figure 10.4.1.

The slopes of the two tangents are

$$y'(x_n, y_n) = f(x_n, y_n), \quad y'(x_{n+1}, y_{n+1}^{(e)}) = f(x_{n+1}, y_{n+1}^{(e)}).$$

Put

$$k_1 = hf(x_n, y_n),$$
$$k_2 = hf(x_{n+1}, y_{n+1}^{(e)}) = hf(x_n + h, y_n + k_1).$$

We now take a step of length h from (x_n, y_n) along the average of the two tangent directions by putting

$$y_{n+1} = y_n + \frac{1}{2}(k_1 + k_2).$$

This method is called **Heun's method**, and it belongs to a large class of **Runge-Kutta methods**.

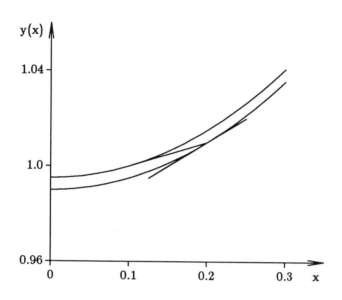

Figure 10.4.1   Two tangent directions.

Example 10.4.2   We solve the initial value problem $y' = xy$, $y(0) = 1$ using Heun's method and step length $h = 0.2$. We first get

$$k_1 = hx_0y_0 = 0.2 \cdot 0 \cdot 1 = 0,$$
$$k_2 = h(x_0 + h)(y_0 + k_1) = 0.2 \cdot 0.2 \cdot 1 = 0.04,$$
$$y_1 = y_0 + 0.5(k_1 + k_2) = 1 + 0.5(0 + 0.04) = 1.02.$$

The second step is

$$k_1 = hx_1y_1 = 0.2 \cdot 0.2 \cdot 1.02 = 0.0408,$$
$$k_2 = h(x_1 + h)(y_1 + k_1) = 0.2 \cdot 0.4 \cdot 1.0608 = 0.084864,$$
$$y_2 = 1.02 + 0.5(0.0408 + 0.084864) = 1.082832.$$

y_2 is an approximation of $y(0.4)$ and the truncation error is approximately $4.55 \cdot 10^{-4}$. Cf. Examples 10.2.2 and 10.3.3, where we used Euler's method.

The price we pay for getting a global truncation error $O(h^2)$ is the evaluation of $f(x, y)$ twice. If we evaluate the function four times, then we can get a method with global truncation error $O(h^4)$.

The classical Runge–Kutta method:

$$k_1 = hf(x_n, y_n),$$
$$k_2 = hf(x_n + h/2, y_n + k_1/2),$$
$$k_3 = hf(x_n + h/2, y_n + k_2/2),$$
$$k_4 = hf(x_n + h, y_n + k_3),$$
$$y_{n+1} = y_n + \frac{1}{6}(k_1 + 2k_2 + 2k_3 + k_4).$$

The method has a global truncation error $O(h^4)$.

This Runge–Kutta method can also be derived geometrically.

Example 10.4.3 If we apply the classical Runge–Kutta method to the initial value problem $y' = xy$, $y(0) = 1$, with step length $h = 0.4$, we get
$$k_1 = hx_0 y_0 = 0,$$
$$k_2 = h(x_0 + h/2)(y_0 + k_1/2) = 0.08,$$
$$k_3 = h(x_0 + h/2)(y_0 + k_2/2) = 0.0832,$$
$$k_4 = h(x_0 + h)(y_0 + k_3) = 0.173312,$$

and

$$y(0.4) \approx y_1 = 1 + \frac{1}{6}(0 + 2 \cdot 0.08 + 2 \cdot 0.0832 + 0.173312) = 1.083285.$$

The truncation error in this approximation is $2 \cdot 10^{-6}$.

It can be shown that the global truncation error in this Runge–Kutta method has an expansion

$$R_T = c_4 h^4 + c_5 h^5 + \dots$$

Therefore, we can estimate the truncation error using Richardson extrapolation by computing approximations of $y(x)$ with two different step lengths, and comparing these.

10.5   An Implicit Method

For certain types of differential equations, the methods we have discussed so far are very inefficient. This applies to problems where there are both very fast and slow processes in the solution. Such problems are called **stiff**. We return to this in Section 10.6. We will now derive a method that is particularly well suited to stiff problems, but it can also be used for differential equations in general.

If we integrate the left hand side of the differential equation

$$y' = f(x, y)$$

over the interval $[x_n, x_{n+1}]$, we get

$$\int_{x_n}^{x_{n+1}} y' \, dx = y(x_{n+1}) - y(x_n).$$

The integral of the right hand side

$$\int_{x_n}^{x_{n+1}} f(x, y(x)) \, dx,$$

cannot be computed directly, since the function $y(x)$ is unknown. Using the trapezoidal rule, we get

$$\int_{x_n}^{x_{n+1}} f(x, y(x)) \, dx \approx \frac{h}{2} \big(f(x_n, y(x_n)) + f(x_{n+1}, y(x_{n+1})) \big).$$

Now, we replace $y(x_n)$ and $y(x_{n+1})$ by the approximations y_n and y_{n+1} everywhere, and we obtain

---

The trapezoidal method:

$$y_{n+1} = y_n + \frac{h}{2} \big(f(x_n, y_n) + f(x_{n+1}, y_{n+1}) \big).$$

The global truncation error is $O(h^2)$.

The methods we have discussed earlier are **explicit**: when we compute y_{n+1}, we have in the right hand side only x_n, y_n and quantities that depend on these. The trapezoidal method is **implicit**: in the right hand side, we have also y_{n+1}. If f is a nonlinear function of y, we must solve a nonlinear equation to get y_{n+1}.

Example 10.5.1   As before, we consider the problem $y' = xy$, $y(0) = 1$. This equation is linear in y. Therefore, we can easily solve for y_{n+1} when we take a step with the trapezoidal method. We use $h = 0.2$, and compute an approximation of $y(0.4)$.

$$y_1 = y_0 + \frac{h}{2}(x_0 y_0 + x_1 y_1) = 1 + 0.1(0 + 0.2y_1),$$

which gives

$$(1 - 0.02)y_1 = 1,$$

and

$$y_1 \approx 1.0204.$$

Similarly,

$$y_2 = y_1 + \frac{h}{2}(x_1 y_1 + x_2 y_2) = 1.0204 + 0.1(0.2 \cdot 1.0204 + 0.4y_2),$$

and

$$y(0.4) \approx y_2 = \frac{1.0408}{(1 - 0.04)} \approx 1.0842.$$

The truncation error in this approximation is $9 \cdot 10^{-4}$.

The equation $y' = e^{-y}$, $y(0) = 1$, is nonlinear in y. The first step with the trapezoidal method and $h = 0.2$ gives

$$y_1 = y_0 + \frac{h}{2}(e^{-y_0} + e^{-y_1}) = 1 + \frac{0.2}{2}(e^{-1} + e^{-y_1}).$$

Thus, we have the equation

$$g(y_1) = y_1 - 0.1e^{-y_1} - (1 + 0.1e^{-1}) = 0,$$

which we can solve using, e.g., Newton–Raphson's method. To get a starting value for the iterations, we use Euler's method: $y_1^{(0)} = y_0 + he^{-y_0} \approx 1.0736$. Two iterations with Newton–Raphson's method give

$$y_1^{(1)} = 1.071053,$$
$$y_1^{(2)} = 1.071053.$$

(Carry out the computations!) We have obtained $y_1 = 1.071053$, and can go on and take another step with the trapezoidal method.

When we use the trapezoidal method, we get the equation

$$y_{n+1} - \frac{h}{2}f(x_{n+1}, y_{n+1}) + A = 0,$$

where A is a known quantity. This equation can be solved using Newton–Raphson's method as in the example above. Note that the derivative that is to be computed in each step is

$$1 - \frac{h}{2}\frac{\partial f}{\partial y}$$

(in the case when the equation is scalar).

Alternatively, y_{n+1} can be computed using the fixed point iteration

$$y_{n+1}^{(0)} = y_n + hf(x_n, y_n),$$
$$y_{n+1}^{(k+1)} = y_n + \frac{h}{2}(f(x_n, y_n) + f(x_{n+1}, y_{n+1}^{(k)})), \qquad k = 0, 1, 2, \ldots.$$

This way of using an implicit method is called a **predictor–corrector** procedure. Here, Euler's method is predictor, which means that it gives the starting value for the fixed point iteration. Then one iterates a couple of times with the trapezoidal method, which is the corrector.

The condition for the iteration to converge is

$$\left|\frac{h}{2}\frac{\partial f}{\partial y}\right| < 1$$

in a neighborhood of (x_{n+1}, y_{n+1}) (cf. Section 4.3).

Example 10.5.2   Assume that we use Euler's method–the trapezoidal method as predictor–corrector to solve the initial value problem $y' = e^{-y}$, $y(0) = 1$. Does the corrector-iteration converge if the step size h is chosen equal to 0.1?

First, we see that the solution of the initial value problem always has a positive derivative, i.e., it is increasing. Therefore, we can never get negative y-values. Then, we have

$$\left|\frac{h}{2}\frac{\partial f}{\partial y}\right| = 0.05e^{-y} < 1.$$

Thus, the corrector-iteration always converges for the chosen step length.

If $|\partial f/\partial y|$ is large, then one must use a very small step length h for the corrector-iteration to converge. In such cases, it is much more efficient to use Newton–Raphson's method.

When solving initial value problems with a predictor–corrector procedure, one very seldom uses Euler's method and the trapezoidal method. It is more efficient to use more accurate methods, e.g., an Adams–Bashforth method as predictor, and an Adams–Moulton method as corrector. These are families of explicit and implicit methods, respectively. The following pair is often used; both methods have global truncation error $O(h^4)$.

$$y_{n+4} = y_{n+3} + \frac{h}{24}(55f_{n+3} - 59f_{n+2} + 37f_{n+1} - 9f_n),$$

$$y_{n+4} = y_{n+3} + \frac{h}{24}(9f_{n+4} + 19f_{n+3} - 5f_{n+2} + f_{n+1}).$$

10.6 Difference Equations

To be able to investigate the stability properties of numerical methods for initial value problems, it is necessary to have some knowledge of difference equations. Such equations arise, e.g., in the discretization of differential equations, and in many respects the theory of difference equations is analogous to that of differential equations. In this section, we will sketch the theory starting from second order difference equations. The generalization to higher order equations is quite obvious, and we give it without proof.

A **linear difference equation of second order** is an equation of the type
$$y_{n+2} + by_{n+1} + cy_n = 0, \qquad n = 0, 1, 2, \ldots.$$
We assume that the equation has **constant coefficients** b and c. Further, we assume that c is nonzero, for otherwise we would have a lower order difference equation. The equation is **homogeneous**, i.e., the right hand side is equal to zero.

Alternatively, one can write a difference equation using difference operators (see Chapter 5) (this explains the name difference equation).

If we give initial values
$$y_0 = \alpha, \qquad y_1 = \beta,$$

then the difference equation uniquely defines a sequence $y_0, y_1, y_2, \ldots$, i.e., we can directly compute $y_2, y_3, \ldots$ from the equation. But, in order to get a general expression for the solution, we try $y_n = r^n$. This leads to
$$r^{n+2} + br^{n+1} + cr^n = 0,$$
or, if we divide by r^n,
$$r^2 + br + c = 0.$$

This is called the **characteristic equation** corresponding to the difference equation. We can now determine the values of r for which $y_n = r^n$ is a solution of the difference equation by solving the characteristic equation. First, assume that the characteristic equation has two simple roots r_1 and r_2. Then, we immediately see that any linear combination $y_n = Ar_1^n + Br_2^n$ is a solution of the difference equation, since

$$y_{n+2} + by_{n+1} + cy_n$$
$$= (Ar_1^{n+2} + Br_2^{n+2}) + b(Ar_1^{n+1} + Br_2^{n+1}) + c(Ar_1^n + Br_2^n)$$
$$= A(r_1^{n+2} + br_1^{n+1} + cr_1^n) + B(r_2^{n+2} + br_2^{n+1} + cr_2^n) = 0.$$

A and B can be determined from the initial values:

$$y_0 = A + B = \alpha, \qquad y_1 = Ar_1 + Br_2 = \beta.$$

Example 10.6.1   The sequence of Fibonacci numbers is defined as

$$y_0 = 0, \qquad y_1 = 1, \qquad y_{n+2} = y_{n+1} + y_n, \qquad n = 0, 1, 2, \ldots.$$

Every number in the sequence is the sum of the two preceding. The sequence is

$$0, 1, 1, 2, 3, 5, 8, 13, 21, 34, \ldots.$$

To derive a general expression for the Fibonacci numbers, we solve the characteristic equation

$$r^2 - r - 1 = 0,$$

which has the roots

$$r_{1,2} = \frac{1 \pm \sqrt{5}}{2}.$$

The general solution is

$$y_n = A\left(\frac{1 + \sqrt{5}}{2}\right)^n + B\left(\frac{1 - \sqrt{5}}{2}\right)^n.$$

Noting that the Fibonacci numbers are integers, this solution may at first sight seem somewhat surprising. But, from the initial values, we determine A and B:

$$y_0 = A + B = 0,$$

$$y_1 = A\frac{1 + \sqrt{5}}{2} + B\frac{1 - \sqrt{5}}{2} = 1,$$

which gives $A = -B = 1/\sqrt{5}$. Thus, the general expression for the Fibonacci numbers is

$$y_n = \frac{1}{\sqrt{5}} \left(\left(\frac{1 + \sqrt{5}}{2} \right)^n - \left(\frac{1 - \sqrt{5}}{2} \right)^n \right).$$

When we use difference equations for the numerical solution of differential equations, we compute every number in the sequence directly from the equation, i.e., we compute $y_{n+2} = -by_{n+1} - cy_n$, $n = 0, 1, \ldots$. The initial values are the input data in this computation. Then it is important to examine the stability of this procedure: if small errors in the input data lead to small errors in the sequence, or if the errors in the input data are grossly magnified.

Example 10.6.2 The difference equation

$$y_{n+2} - 2.5y_{n+1} + y_n = 0$$

has the general solution

$$y_n = A\, 2^{-n} + B\, 2^n.$$

With the initial values

$$y_0 = 2, \quad y_1 = 1,$$

the solution is $y_n = 2^{-n+1}$, which is a decreasing sequence. If we perturb the input data somewhat

$$y_0 = 2, \quad y_1 = 1 + \epsilon,$$

we get the solution

$$y_n = \left(2 - \frac{2\epsilon}{3} \right) 2^{-n} + \frac{2\epsilon}{3} 2^n,$$

which, for n large enough, is an increasing sequence.

Obviously, the problem of computing the sequence from the difference equation and the given initial values is very sensitive to perturbations in the input data. Similar problems can occur if a differential equation is discretized using a bad method; see Section 10.7.

If the characteristic equation of a difference equation has a **double root** r_1, then $y_n = nr_1^{n-1}$ is a solution of the difference equation, since

$$y_{n+2} + by_{n+1} + cy_n = (n+2)r_1^{n+1} + b(n+1)r_1^n + cnr_1^{n-1}$$
$$= n(r_1^2 + br_1 + c)r_1^{n-1} + (2r_1 + b)r_1^n = 0.$$

(The last equality is valid due to the fact that a double root of the second degree equation $p(x) = 0$ is also a root of the equation $p'(x) = 0$.) Therefore, the general solution of the difference equation is $y_n = Anr_1^{n-1} + Br_1^n$.

The solution of an **inhomogeneous** difference equation is equal to the sum of the general solution of the *homogeneous* equation and a **particular solution** of the inhomogeneous equation. If the right hand side is a polynomial in n, then one can, in certain cases, find a particular solution quite easily.

Example 10.6.3   Given the inhomogeneous equation

$$y_{n+2} - 3y_{n+1} + 2y_n = n.$$

The corresponding homogeneous equation has the general solution

$$y_n = A\,1^n + B\,2^n.$$

To find a particular solution of the inhomogeneous equation, we assume that y_n is a polynomial in n. Since the sum of the coefficients of the left hand side is equal to zero, the degree of the polynomial y_n is decreased if it is inserted into the left hand side (check that this is valid for an arbitrary polynomial). We therefore try a second degree polynomial

$$y_n = \alpha n^2 + \beta n.$$

Inserting this into the difference equation and identifying coefficients, we get $\alpha = \beta = -\frac{1}{2}$. Thus, the general solution of the inhomogeneous equation is

$$y_n = A\,1^n + B\,2^n - \frac{1}{2}(n^2 + n).$$

We now summarize the results and formulate them for a difference equation of order k.

A linear, homogeneous difference equation of order k is an equation of the type

$$y_{n+k} + a_1 y_{n+k-1} + \ldots + a_k y_n = 0, \qquad n = 0, 1, 2, \ldots.$$

If $r_1, r_2, \ldots, r_k$ are simple roots of the characteristic equation

$$r^k + a_1 r^{k-1} + \ldots + a_k = 0,$$

then the difference equation has the general solution

$$y_n = c_1 r_1^n + c_2 r_2^n + \ldots + c_k r_k^n,$$

where $c_1, c_2, \ldots, c_k$ can be determined from k initial values.

If r_i is a root of multiplicity m of the characteristic equation, then

$$y_n = p_{m-1}(n) r_i^n,$$

where p_{m-1} is a polynomial of degree $m - 1$, satisfies the difference equation.

10.7 Stability

When an initial value problem for an ordinary differential equation is discretized, one gets a difference equation from which a sequence of approximations to the solution of the differential equation is computed. It is necessary to analyze: a) how well the difference equation approximates the differential equation (truncation error), b) what happens when the step length h tends to zero (convergence), and c) how sensitive the difference equation is to perturbations in the data (stability).

In this section, we will study the stability of some numerical methods by investigating their behavior when applied to the **test problem**

$$y' = \lambda y, \quad y(0) = 1,$$

where λ is negative (more generally, one often lets λ be complex with negative real part). The analytic solution $y(x) = e^{\lambda x}$ is decreasing, and it is reasonable to require that the numerical method gives a decreasing sequence.

We mentioned earlier that one may consider discretizing a differential equation by replacing the derivative by a central difference, instead of a forward difference as in Euler's method. One then gets

$$\frac{y(x_{n+1}) - y(x_{n-1})}{2h} \approx f(x_n, y(x_n)),$$

and the recursion formula for the **midpoint method** is

$$y_{n+1} = y_{n-1} + 2hf(x_n, y_n).$$

Note that, in contrast to the previous methods, this is a **two-step method**: to compute y_{n+1}, we must use both y_n and y_{n-1}. We first show that this method, applied to a simple equation, gives a solution with growing oscillations. Then, we will investigate the method theoretically.

Example 10.7.1   The initial value problem $y' = -2y$, $y(0) = 1$, has the solution $y(x) = e^{-2x}$. Applying the midpoint method to this equation, we get

$$y_{n+1} = y_{n-1} + 2h(-2y_n).$$

The initial value is $y_0 = 1$. To use the method, we must also compute an approximation to y_1, which we do using Euler's method. It gives $y_1 = y_0 + h(-2y_0)$. With $h = 0.1$, we get

x_n	y_n	$y(x_n)$	Error
0.1	0.8000	0.8187	$-1.87 \cdot 10^{-2}$
0.2	0.6800	0.6703	$0.97 \cdot 10^{-2}$
0.3	0.5280	0.5488	$-2.08 \cdot 10^{-2}$
0.4	0.4688	0.4493	$1.95 \cdot 10^{-2}$
0.5	0.3405	0.3679	$-2.74 \cdot 10^{-2}$
0.6	0.3326	0.3012	$3.14 \cdot 10^{-2}$
0.7	0.2074	0.2466	$-3.92 \cdot 10^{-2}$
0.8	0.2496	0.2019	$4.77 \cdot 10^{-2}$
0.9	0.1076	0.1653	$-5.77 \cdot 10^{-2}$
1.0	0.2066	0.1353	$7.13 \cdot 10^{-2}$

We see that very soon the numerical solution starts to oscillate. After $x = 0.7$, it is no longer monotonically decreasing. If we continue, then the growing oscillation will soon dominate the solution completely. Obviously, the midpoint method does not work at all for this equation.

The solution is illustrated in the following Figure 10.7.2.

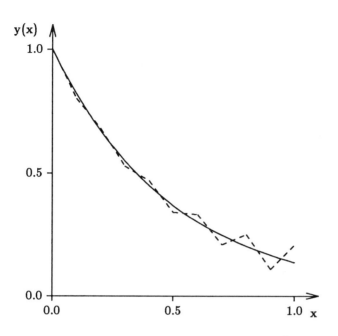

Figure 10.7.2   The midpoint method applied to $y' = -2y$, $y(0) = 1$.

If we apply the midpoint method to the test problem $y' = \lambda y$, $y(0) = 1$, we get the difference equation

$$y_{n+1} = y_{n-1} + 2h\lambda y_n,$$

and the characteristic equation

$$r^2 - 2h\lambda r - 1 = 0,$$

which has the roots

$$r_{1,2} = h\lambda \pm \sqrt{h^2\lambda^2 + 1}\,.$$

If $h^2\lambda^2 < 1$, we can use the series expansion

$$\sqrt{1 + h^2\lambda^2} = 1 + \frac{h^2\lambda^2}{2} - \frac{h^4\lambda^4}{8} + \dots,$$

and we get

$$r_1 = 1 + h\lambda + \frac{h^2\lambda^2}{2} - \frac{h^4\lambda^4}{8} + \dots = e^{h\lambda} + O(h^3),$$

$$r_2 = -1 + h\lambda - \frac{h^2\lambda^2}{2} + \frac{h^4\lambda^4}{8} + \dots = -e^{-h\lambda} + O(h^3).$$

The general solution of the difference equation is

$$y_n = c_1 r_1^n + c_2 r_2^n = c_1 e^{nh\lambda} + c_2(-1)^n e^{-nh\lambda} + O(h^2).$$

Since $x_n = nh$, we now have

$$y_n = c_1 e^{x_n\lambda} + c_2(-1)^n e^{-x_n\lambda} + O(h^2).$$

The initial values $y_0 = 1$, $y_1 = e^{h\lambda}$ give

$$c_1 = 1 - c_2, \qquad c_2 = \frac{e^{h\lambda} - r_1}{r_2 - r_1}.$$

If the step length h is small, then $c_1 \approx 1$, $c_2 \approx 0$, but c_2 is nonzero.

We thus see that the midpoint method produces a solution with two components: the first corresponds to the exact solution of the differential equation, while the other is an oscillating spurious solution.

Since $\lambda < 0$, the exact solution is decreasing. The spurious solution is increasing with x (cf. Example 10.7.1) and, eventually, it will dominate the numerical solution. Even if the initial values are chosen so that $c_2 = 0$, the spurious solution will be present in the numerical solution due to rounding errors (essentially, it is impossible to make c_2 exactly equal to zero in floating point arithmetic).

The analysis shows that the midpoint method is **unstable**.

Euler's method applied to the test equation gives

$$y_{n+1} = y_n + h\lambda y_n = (1 + h\lambda)y_n.$$

This first order difference equation has the solution

$$y_n = (1 + h\lambda)^n y_0,$$

and we see that the solution is decreasing if

$$|1 + h\lambda| < 1,$$

or, equivalently,

$$-2 < h\lambda < 0.$$

With negative λ, we must therefore choose the step length h so that

$$h < \frac{2}{|\lambda|}.$$

If $|\lambda|$ is large, then we must choose h very small in Euler's method to obtain a decreasing solution.

The trapezoidal method applied to the test equation gives

$$y_{n+1} = y_n + \frac{h}{2}\lambda(y_n + y_{n+1}),$$

or, equivalently,

$$y_{n+1} = \frac{1 + h\lambda/2}{1 - h\lambda/2} y_n.$$

Here, we have

$$\left| \frac{1 + h\lambda/2}{1 - h\lambda/2} \right| < 1,$$

for negative λ, and for all values of h. The trapezoidal method always gives decreasing solutions for negative λ.

It is this stability property that makes the trapezoidal method useful for **stiff** differential equations (or stiff systems of differential equations). Typical of such equations is the existence of some components in the solution that decrease very fast, and some components that decrease quite slowly. The former correspond to negative values of λ of large magnitude in the test equation, and the latter to negative λ of relatively small magnitude.

Consider the initial value problem

$$y'' + 101y' + 100y = 0, \qquad y(0) = 1.01, \qquad y'(0) = -2.$$

This can be written as a system

$$\begin{pmatrix} y_1' \\ y_2' \end{pmatrix} = \begin{pmatrix} 0 & 1 \\ -100 & -101 \end{pmatrix} \begin{pmatrix} y_1 \\ y_2 \end{pmatrix}, \qquad \begin{pmatrix} y_1(0) \\ y_2(0) \end{pmatrix} = \begin{pmatrix} 1.01 \\ -2 \end{pmatrix}.$$

The matrix of this system has the eigenvalues -100 and -1, which correspond to $\lambda = -100$ and $\lambda = -1$, respectively, in the test equation. The solution is

$$y_1(x) = y(x) = 0.01e^{-100x} + e^{-x}.$$

The first term in the solution decays very rapidly: at $x = 0.1$, it is of the order of magnitude 10^{-7}. When one solves such a system over a large interval, one would like to take large steps as soon as the rapidly decaying components have disappeared. This cannot be done with explicit methods like Euler's method or a Runge–Kutta method because of their stability properties. But, with an implicit method, it is possible to do so.

Example 10.7.3   We solve the stiff problem above using Euler's method and the trapezoidal method (applied to the equation written as a system). We start at $x = 0.1$, and take the analytic solution as initial value in this point. We use the step length $h = 0.1$. The result is

x_n	Euler	Trapezoidal	Analytic
0.2	0.8143	0.8187	0.8187
0.3	0.7330	0.7407	0.7408
0.4	0.6593	0.6702	0.6703
0.5	0.5966	0.6063	0.6065
0.6	0.5075	0.5486	0.5488
0.7	0.7221	0.4963	0.4966
0.8	−1.7390	0.4491	0.4493
0.9	19.9359	0.4063	0.4066
1.0	−175.5667	0.3676	0.3679

According to the analysis of Euler's method, one would need to use the step length

$$h < \frac{2}{100} = 0.02$$

in order to get a decreasing numerical solution.

In stiff systems, it is not unusual that the rapid and the slow components correspond to $\lambda = -10^4$ and $\lambda = -1$, respectively. Here, it is necessary to use implicit methods.

We summarize this section: When the midpoint method is applied to the test problem (with negative λ) we get an oscillating solution whose magnitude grows exponentially. With Euler's method (and other explicit methods), we get a decreasing solution if we choose h small enough. The trapezoidal method gives a decreasing solution for all values of h.

10.8   Boundary Value Problems

Many physical processes can be described using second order differential equations, with extra conditions given in two points. The solution is required in the interval between the two boundary points. Often such a problem can be written in the form

$$y'' = f(x, y, y'), \qquad y(a) = \alpha, \qquad y(b) = \beta,$$

where f is a given function. This is called a **boundary value problem** for an ordinary differential equation. Remember that two extra conditions are needed for the solution of a second order differential equation to be determined. If they are given at the same point, then we have an initial value problem, and if they are given in two points we have a boundary value problem. The extra conditions are called **boundary values**.

Example 10.8.1   Assume that we have a thin rod of length 1 with variable heat conduction properties (e.g., different parts of the rod may be made of different materials). Further, assume that the rod is insulated along its length, and that the endpoints are kept at different constant temperatures. If the experiment is carried out long enough, then the temperature of the rod will be stationary, i.e., independent of time. Let $y(x)$ denote the temperature of the rod at the point x. Then y satisfies the differential equation

$$\frac{d}{dx}(k(x)\frac{dy}{dx}) = 0,$$

with boundary values

$$y(0) = \alpha, \qquad y(1) = \beta.$$

$k(x)$ is the heat conduction coefficient of the rod, and α and β denote the temperature at the endpoints.

If, furthermore, the rod contains a heat source, e.g., an electrical resistance or a radioactive isotope, then the equation is inhomogeneous:

$$\frac{d}{dx}(k(x)\frac{dy}{dx}) = f(x),$$

where $f(x)$ describes the production of heat as a function of x.

If k and f are constant, then the boundary value problem can be solved analytically (e.g., if k is constant and $f(x) = 0$, then the solution is a straight line). For problems with non-constant coefficients, it is usually necessary to discretize and compute an approximate solution.

In Chapter 1, we gave an example, from structural mechanics, of a boundary value problem for a fourth order differential equation.

In the following sections, we will discuss three different methods: the (finite) difference method, a finite element method, and the shooting method. For simplicity, we will present the theory using a special case, e.g., the linear equation

$$y'' - q(x)y = f(x).$$

10.9   The Difference Method

The difference method is based on dividing the interval $[a, b]$ into subintervals, which here we assume all have length

$$h = \frac{b - a}{N}.$$

Put

$$x_n = a + nh, \qquad n = 0, 1, 2, \ldots, N,$$

and let y_n be an approximation of $y(x_n)$.

Figure 10.9.1   The interval $[a, b]$ divided into N subintervals.

Note that we already know y_0 and y_N, since they are given by the boundary values.

Consider the special case when the equation is

$$y'' - q(x)y = f(x), \qquad q(x) \geq 0.$$

We now discretize the differential equation by replacing the second derivative by a difference approximation at the point $x = x_n$ (cf. Chapter 6):

$$y''(x_n) \approx \frac{y_{n+1} - 2y_n + y_{n-1}}{h^2}.$$

We can do this at all interior points x_n, $n = 1, 2, \ldots N - 1$. When inserting this into the differential equation, we get

$$\frac{y_{n+1} - 2y_n + y_{n-1}}{h^2} - q(x_n)y_n = f(x_n), \qquad n = 1, 2, \ldots, N - 1,$$

or, equivalently,

$$y_{n+1} - (2 + h^2 q_n)y_n + y_{n-1} = h^2 f_n, \qquad n = 1, 2, \ldots, N - 1,$$

where we have used q_n and f_n as notations for $q(x_n)$ and $f(x_n)$.

We must look more closely at the first and last equations, since y_0 and y_N enter there. For $n = 1$, we have

$$y_2 - (2 + h^2 q_1)y_1 + y_0 = h^2 f_1,$$

which, using the boundary value $y_0 = \alpha$, we can write

$$y_2 - (2 + h^2 q_1)y_1 = h^2 f_1 - \alpha.$$

Analogously, we get, for $n = N - 1$,

$$-(2 + h^2 q_{N-1})y_{N-1} + y_{N-2} = h^2 f_{N-1} - \beta.$$

Now, we assemble the unknowns y_n, $n = 1, 2, \ldots, N - 1$, into a vector:

$$y = \begin{pmatrix} y_1 \\ y_2 \\ \vdots \\ y_{N-2} \\ y_{N-1} \end{pmatrix}.$$

Then, we can write the equations as a linear system

$$Ay = b,$$

where the matrix A is tridiagonal:

$$\begin{pmatrix} -(2 + h^2 q_1) & 1 & & & & \\ 1 & -(2 + h^2 q_2) & 1 & & & \\ & \ddots & \ddots & \ddots & & \\ & & 1 & -(2 + h^2 q_{N-2}) & 1 & \\ & & & 1 & -(2 + h^2 q_{N-1}) \end{pmatrix};$$

and the right hand side is given by

$$b = \begin{pmatrix} h^2 f_1 - \alpha \\ h^2 f_2 \\ \vdots \\ h^2 f_{N-2} \\ h^2 f_{N-1} - \beta \end{pmatrix}.$$

We refer to this method of discretizing the boundary value problem as the **difference method**, since the derivatives are replaced by differences. Sometimes, the method is called the *finite* difference method. Another name is the **band matrix method**: when the method is applied to a linear differential equation, then one gets a tridiagonal system of linear equations.

We have made the assumption that $q(x) \geq 0$. This will ensure that the matrix A is diagonally dominant (cf. Section 8.4). A diagonally dominant matrix is nonsingular, and the linear system of equations has a unique solution for any right hand side b. When solving the system, one can use Gaussian elimination without pivoting, and the operation count is $O(N)$ for large N (see Section 8.8).

Example 10.9.2   The boundary value problem

$$y'' - y = 0, \qquad y(0) = 0, \qquad y(1) = \sinh(1),$$

has the analytic solution $y(x) = \sinh(x)$. We discretize the problem using the difference method with $h = 0.25$.

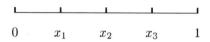

$$0 \qquad x_1 \qquad x_2 \qquad x_3 \qquad 1$$

The boundary values give $y_0 = 0$ and $y_4 = \sinh(1)$. The unknowns y_1, y_2, y_3 satisfy

$$y_{n+1} - (2 + h^2)y_n + y_{n-1} = 0, \qquad n = 1, 2, 3,$$

or, in matrix form $Ay = b$, where

$$A = \begin{pmatrix} -(2+h^2) & 1 & 0 \\ 1 & -(2+h^2) & 1 \\ 0 & 1 & -(2+h^2) \end{pmatrix},$$

and the right hand side is

$$b = \begin{pmatrix} 0 \\ 0 \\ -\sinh(1) \end{pmatrix}.$$

The solution of the linear system is given in the table.

x_n	y_n	Error
0.25	0.252803	$1.9 \cdot 10^{-4}$
0.50	0.521406	$3.1 \cdot 10^{-4}$
0.75	0.822598	$2.8 \cdot 10^{-4}$

If we halve the step length, we get

x_n	y_n	Error
0.125	0.125351	$2.5 \cdot 10^{-5}$
0.250	0.252660	$4.8 \cdot 10^{-5}$
0.375	0.383918	$6.7 \cdot 10^{-5}$
0.500	0.521174	$7.8 \cdot 10^{-5}$
0.625	0.666573	$8.1 \cdot 10^{-5}$
0.750	0.822387	$7.1 \cdot 10^{-5}$
0.875	0.991052	$4.5 \cdot 10^{-5}$

When we halved the step length, the error was reduced by a factor of four approximately. This indicates that the truncation error of the method is $O(h^2)$.

The following theorem, which we give without proof, shows that the truncation error of the difference method is indeed $O(h^2)$.

Theorem 10.9.3   Assume that $q(x) \geq 0$, and let the boundary value problem

$$y'' - q(x)y = f(x), \qquad y(a) = \alpha, \qquad y(b) = \beta,$$

have the solution $y(x)$. Further, assume that $|y^{(4)}(x)| \leq M$ for all $x \in [a, b]$. Let y_n denote the approximation of $y(x_n)$ obtained with the difference method. Then

$$|y(x_n) - y_n| \leq \frac{Mh^2}{24}(x_n - a)(b - x_n), \qquad n = 1, 2, \ldots, N - 1.$$

From the theorem, we see that we can use Richardson extrapolation to refine the approximate solution.

Example 10.9.4   In the previous example, we can perform Richardson extrapolation for $x = 0.25$, 0.5 and 0.75. As we have halved the step, the denominator of the first correction term is $\Delta/3$. The result is

x_n	y_{extr}	Error
0.25	0.25261233	$1 \cdot 10^{-8}$
0.50	0.52109667	$1.4 \cdot 10^{-6}$
0.75	0.82231667	$6 \cdot 10^{-8}$

We have discussed the difference method applied to the special case of a linear differential equation without a first derivative. It is easy to generalize to problems with a first derivative, but, in this case, it may happen that the matrix is not diagonally dominant if the step length is chosen too large (see the exercises).

If the difference method is applied to the problem

$$y'' = f(x, y), \qquad y(a) = \alpha, \qquad y(b) = \beta,$$

where $f(x, y)$ is nonlinear in y, then one gets

$$y_{n+1} - 2y_n + y_{n-1} = h^2 f(x_n, y_n), \qquad n = 1, 2, \ldots, N - 1.$$

This is a nonlinear system

$$Ay = F(x, y),$$

where A is the tridiagonal matrix

$$A = \begin{pmatrix} -2 & 1 & & & \\ 1 & -2 & 1 & & \\ & \ddots & \ddots & \ddots & \\ & & 1 & -2 & 1 \\ & & & 1 & -2 \end{pmatrix},$$

and the vector valued function F is given by

$$F(x, y) = \begin{pmatrix} h^2 f(x_1, y_1) - \alpha \\ h^2 f(x_2, y_2) \\ \vdots \\ h^2 f(x_{N-2}, y_{N-2}) \\ h^2 f(x_{N-1}, y_{N-1}) - \beta \end{pmatrix}.$$

The nonlinear system can be solved using fixed point iteration (cf. Section 4.7):

$$Ay^{(k+1)} = F(x, y^{(k)}).$$

In each iteration, one then has to solve a tridiagonal system with the matrix A.

It can be shown that, under quite mild restrictions on the function f, the truncation error is $O(h^2)$ in the nonlinear case also.

10.10   A Finite Element Method

Finite element methods are often used for solving problems in structural mechanics. They are based on the principle of subdividing the construction under study into small parts, "finite elements"; it is possible to formulate equations for how each part is influenced when a load is put on the construction. Each part depends on its neighbors, and this leads to a system of equations that describes the total effect on the construction.

In structural mechanics computations, it is usually *partial* differential equations that are solved. Often the **Rayleigh–Ritz method** is used. This method is closely related to the physical background, since it is based on the principle of minimizing the potential energy of the system. Here we are going to describe a similar method, **Galerkin's method**, applied to a boundary value problem for an *ordinary* differential equation. When this method is used as we do here, it gives exactly the same result as the Rayleigh–Ritz method.

It is not our aim to give a full account of finite element methods. Instead, we want to introduce some mathematical and numerical ideas that the method is based on. We deliberately avoid certain details, and therefore the presentation is not completely stringent.

To formulate Galerkin's method, we define the scalar product of two functions h and g defined in the interval $[a, b]$ (cf. Chapter 9):

$$(h, g) = \int_a^b h(x)g(x)\,dx.$$

Two functions h and g are said to be **orthogonal** if $(h, g) = 0$.

Consider the boundary value problem

$$Ly = -y'' + qy = f, \qquad y(a) = y(b) = 0,$$

where we assume that $q \geq 0$ is a constant, and that $f(x)$ is not identically zero. We introduce the notation L for the differential operator which maps a twice differentiable function y onto $-y'' + qy$.

Let V be a class of **test functions** that satisfy the boundary conditions:

$$V = \{v \mid v' \text{ is piecewise continuous and bounded on } [a, b],$$
$$\text{and } v(a) = v(b) = 0\}.$$

If y satisfies the differential equation $Ly = f$, then, trivially,

$$(Ly - f, v) = 0,$$

for all $v \in V$. Conversely, it can be shown that if y is such that

$$(Ly, v) = (f, v),$$

for all $v \in V$, then y also satisfies $Ly = f$. The equation $(Ly, v) = (f, v)$ is called the weak form of the differential equation.

The weak form:
$$(Ly, v) = (f, v),$$

for all $v \in V$.

The weak form is similar to the normal equations for an overdetermined linear system of equations; there, we define a residual vector and require it to be orthogonal to all vectors in a certain linear space. In the weak form, the residual function $r = Ly - f$ is required to be orthogonal to all functions in V.

We will now reformulate the left hand side in the weak form. The definitions of the operator L and the scalar product give

$$(Ly, v) = (-y'' + qy, v) = -\int_a^b y''(x)v(x)\,dx + q\int_a^b y(x)v(x)\,dx.$$

We perform a partial integration of the first term, and get

$$(Ly, v) = \left[-y'(x)v(x)\right]_a^b + \int_a^b y'(x)v'(x)\,dx + q\int_a^b y(x)v(x)\,dx$$
$$= \int_a^b y'(x)v'(x)\,dx + q\int_a^b y(x)v(x)\,dx;$$

since we have assumed that the functions $v \in V$ satisfy the boundary conditions, the integrated term is equal to zero. Thus, the weak form $(Ly, v) = (f, v)$ can be written

---

The weak form **partially integrated**:

$$(Ly, v) = (y', v') + q(y, v) = (f, v).$$

By a partial integration, we have moved one derivative from y to v.

Let y^h denote an approximate solution of the boundary value problem. We write y^h in the form

$$y^h = \sum_{j=1}^{N-1} c_j \varphi_j,$$

where φ_j, $j = 1, 2, \ldots, N-1$, are given functions. The notation y^h indicates that they are related to a discretization with step length h, but, for a while, we refrain from specifying them. We merely assume that the functions φ_j are linearly independent (cf. Chapter 9); i.e., they constitute a basis in an $(N-1)$-dimensional space of functions that we call V^h. Further, we assume that all the basis functions φ_j satisfy the boundary conditions:

$$\varphi_j(a) = \varphi_j(b) = 0, \qquad j = 1, 2, \ldots, N-1.$$

Inserting the expression for y^h into the weak form, we get

$$(Ly^h, v) = (f, v).$$

We can also "discretize" the functions v by requiring them to belong to V^h. We get

$$(Ly^h, v^h) = (f, v^h),$$

for all $v^h \in V^h$. This is a special case of Galerkin's method.

Let V^h be a finite-dimensional class of functions with basis $(\varphi_i)_{i=1}^{N-1}$. **Galerkin's method** applied to the equation $Ly = f$ amounts to determining the function $y^h \in V^h$ that satisfies

$$(Ly^h, v^h) = (f, v^h),$$

for all functions $v^h \in V^h$.

From the assumption that all the functions φ_i satisfy the boundary conditions, we see that also y^h will do so.

The requirement that $(Ly^h, v^h) = (f, v^h)$ holds for all $v^h \in V^h$ is equivalent to the same requirement with v^h replaced by all the basis functions in V^h:

$$(Ly^h, \varphi_i) = (f, \varphi_i), \qquad i = 1, 2, \ldots, N-1.$$

We will now show that this is a linear system of equations for the coefficients c_j. From $y^h = \sum_{j=1}^{N-1} c_j \varphi_j$, we get, using the partially integrated form,

$$(Ly^h, \varphi_i) = \sum_{j=1}^{N-1} c_j(\varphi_j', \varphi_i') + q \sum_{j=1}^{N-1} c_j(\varphi_j, \varphi_i) = (f, \varphi_i), \quad i = 1, 2, \ldots, N-1.$$

This is a linear system of equations for the coefficients c_j:

$$Kc = F.$$

The matrix K is the sum of two matrices

$$K = K_1 + K_0,$$

where the **stiffness matrix** K_1 is given by

$$K_1 = \begin{pmatrix} (\varphi_1', \varphi_1') & (\varphi_1', \varphi_2') & \cdots & (\varphi_1', \varphi_{N-1}') \\ (\varphi_2', \varphi_1') & (\varphi_2', \varphi_2') & \cdots & (\varphi_2', \varphi_{N-1}') \\ \vdots & \vdots & & \vdots \\ (\varphi_{N-1}', \varphi_1') & (\varphi_{N-1}', \varphi_2') & \cdots & (\varphi_{N-1}', \varphi_{N-1}') \end{pmatrix},$$

the **mass matrix** K_0 is

$$K_0 = q \begin{pmatrix} (\varphi_1, \varphi_1) & (\varphi_1, \varphi_2) & \cdots & (\varphi_1, \varphi_{N-1}) \\ (\varphi_2, \varphi_1) & (\varphi_2, \varphi_2) & \cdots & (\varphi_2, \varphi_{N-1}) \\ \vdots & \vdots & & \vdots \\ (\varphi_{N-1}, \varphi_1) & (\varphi_{N-1}, \varphi_2) & \cdots & (\varphi_{N-1}, \varphi_{N-1}) \end{pmatrix},$$

and the right hand side F is

$$F_i = (f, \varphi_i), \qquad i = 1, 2, \ldots, N-1.$$

Note that the matrix K is *symmetric*, since both K_1 and K_0 are. Further, one can show that K is *positive definite* (under the given assumptions on the operator L).

So far, our presentation has been independent of the choice of basis functions φ_j. Now, we will choose **linear elements** as basis functions. First, we divide the interval $[a, b]$ into N subintervals, for simplicity equidistant:

$$x_j = a + jh, \qquad h = \frac{b - a}{N}, \quad j = 0, 1, \ldots, N.$$

The linear element φ_j is a piecewise linear function, defined by

$$\varphi_j(x) = \begin{cases} 0, & x_0 \leq x \leq x_{j-1}, \\ \dfrac{x - x_{j-1}}{h}, & x_{j-1} \leq x \leq x_j, \\ \dfrac{x_{j+1} - x}{h}, & x_j \leq x \leq x_{j+1}, \\ 0, & x_{j+1} \leq x \leq x_N. \end{cases}$$

In particular, we have $\varphi_j(x_j) = 1$. The linear elements φ_{j-1} and φ_j are illustrated in Figure 10.10.1.

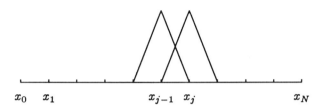

Figure 10.10.1   The linear elements φ_{j-1} and φ_j.

The functions φ_j were introduced already in Section 5.10 . In the present chapter, we call them linear elements, but there they were called linear B-splines. The function space V^h that is spanned by the functions φ_j is therefore a space of linear splines, i.e., piecewise straight lines.

Note that φ_j is equal to zero except in the two intervals next to the point x_j. Therefore, most of the scalar products in the stiffness and mass matrices are equal to zero. More precisely,

$$(\varphi_j', \varphi_k') = (\varphi_j, \varphi_k) = 0, \qquad |j - k| \geq 2.$$

In other words: Both matrices are *tridiagonal*. We now compute the nonzero elements. The diagonal elements of the stiffness matrix are

$$(\varphi_j', \varphi_j') = \int_{x_{j-1}}^{x_j} \left(\frac{1}{h}\right)^2 dx + \int_{x_j}^{x_{j+1}} \left(-\frac{1}{h}\right)^2 dx = \frac{2}{h}.$$

Since $[x_{j-1}, x_j]$ is the only interval where φ_{j-1} and φ_j are simultaneously nonzero, the elements next to the main diagonal are

$$(\varphi_{j-1}', \varphi_j') = \int_{x_{j-1}}^{x_j} \left(\frac{1}{h}\right)\left(-\frac{1}{h}\right) dx = -\frac{1}{h}.$$

Thus, the stiffness matrix is

$$K_1 = \frac{1}{h} \begin{pmatrix} 2 & -1 & & & \\ -1 & 2 & -1 & & \\ & \ddots & \ddots & \ddots & \\ & & -1 & 2 & -1 \\ & & & -1 & 2 \end{pmatrix}.$$

(Cf. the difference method.)

Similarly, we get the diagonal elements of the mass matrix

$$q(\varphi_j, \varphi_j) = q \int_{x_{j-1}}^{x_j} \left(\frac{x - x_{j-1}}{h} \right)^2 dx + q \int_{x_j}^{x_{j+1}} \left(\frac{x_{j+1} - x}{h} \right)^2 dx = q\frac{2h}{3}.$$

The elements next to the main diagonal are

$$q(\varphi_{j-1}, \varphi_j) = q \int_{x_{j-1}}^{x_j} \frac{x_j - x}{h} \frac{x - x_{j-1}}{h} dx = q\frac{h}{6}.$$

Thus, the mass matrix is

$$K_0 = \frac{qh}{6} \begin{pmatrix} 4 & 1 & & & \\ 1 & 4 & 1 & & \\ & \ddots & \ddots & \ddots & \\ & & 1 & 4 & 1 \\ & & & 1 & 4 \end{pmatrix}.$$

Example 10.10.2   We discretize the boundary value problem

$$-y'' + y = 1, \qquad y(0) = y(1) = 0,$$

using linear elements and the step length $h = 0.25$.

The stiffness and mass matrices are

$$K_1 = \frac{1}{0.25} \begin{pmatrix} 2 & -1 & 0 \\ -1 & 2 & -1 \\ 0 & -1 & 2 \end{pmatrix}, \qquad K_0 = \frac{0.25}{6} \begin{pmatrix} 4 & 1 & 4 \\ 1 & 4 & 1 \\ 4 & 1 & 4 \end{pmatrix}.$$

The components of the right hand side are given by

$$F_i = (f, \varphi_i) = \int_0^1 \varphi_i(x)\, dx = h = 0.25, \qquad i = 1, 2, 3.$$

The finite element equation is (we have multiplied the left and right hand sides by h)

$$\begin{pmatrix} 2 + 1/24 & -1 + 1/96 & 0 \\ -1 + 1/96 & 2 + 1/24 & -1 + 1/96 \\ 0 & -1 + 1/96 & 2 + 1/24 \end{pmatrix} \begin{pmatrix} c_1 \\ c_2 \\ c_3 \end{pmatrix} = \begin{pmatrix} 1/16 \\ 1/16 \\ 1/16 \end{pmatrix},$$

which has the solution

$$\begin{pmatrix} c_1 \\ c_2 \\ c_3 \end{pmatrix} \approx \begin{pmatrix} 0.0857 \\ 0.1137 \\ 0.0857 \end{pmatrix}.$$

In Figure 10.10.3, we have plotted the finite element solution y^h for two different step lengths.

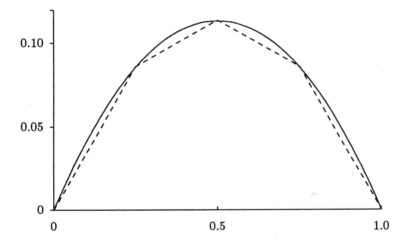

Figure 10.10.3 The finite element solution y^h for $h = 0.25$ (dashed) and $h = 0.01$ (solid).

The following can be shown.

Theorem 10.10.4 The piecewise linear finite element approximation y^h, derived as in this section, satisfies

$$\|y - y^h\| \le Ch^2 \|f\|,$$

for some constant C. (The norm is defined as $\|y\| = \sqrt{(y,y)}$.)

We summarize:

The finite element equation for the boundary value problem

$$-y'' + qy = f, \qquad y(a) = y(b) = 0,$$

discretized using linear elements φ_j on an equidistant partitioning of the interval, is given by

$$Kc = F, \qquad K = K_1 + K_0,$$

where the stiffness matrix K_1 is

$$K_1 = \frac{1}{h} \begin{pmatrix} 2 & -1 & & & \\ -1 & 2 & -1 & & \\ & \ddots & \ddots & \ddots & \\ & & -1 & 2 & -1 \\ & & & -1 & 2 \end{pmatrix},$$

the mass matrix K_0 is

$$K_0 = q\frac{h}{6} \begin{pmatrix} 4 & 1 & & & \\ 1 & 4 & 1 & & \\ & \ddots & \ddots & \ddots & \\ & & 1 & 4 & 1 \\ & & & 1 & 4 \end{pmatrix},$$

and the right hand side F is defined as

$$F_i = (f, \varphi_i), \qquad i = 1, 2, \ldots, N-1.$$

The approximate solution y^h is a piecewise linear function

$$y^h = \sum_{j=1}^{N-1} c_j \varphi_j.$$

In the same way as we introduced cubic splines in Chapter 5, we can derive finite element methods for cubic elements. This leads to finite element equations with band width 5, and the truncation error will be $O(h^4)$.

Finite element methods applied to boundary value problems for ordinary differential equations have no special advantages over the difference method. However, for the solution of partial differential equations, finite element methods are much more flexible. The perhaps greatest advantage is the ease with which it is possible to treat complicated geometries and different boundary conditions. Further, there is a well-developed mathematical theory, which allows one to prove convergence and derive error estimates, even for complicated problems.

10.11   The Shooting Method

We will now study a third method for solving a boundary value problem

$$y'' = f(x, y, y'), \qquad y(a) = \alpha, \qquad y(b) = \beta.$$

We assume that the problem has a unique solution. Further, we must assume that the initial value problem for the same equation has a unique solution.

Suppose that we know the derivative of the solution at the left endpoint of the interval; i.e., we know that $y'(a) = \gamma$, for some γ. Then, we have an *initial value problem*

$$y'' = f(x, y, y'), \qquad y(a) = \alpha, \qquad y'(a) = \gamma,$$

which we can also write as a system of first order differential equations (put $y' = v$)

$$\begin{cases} y' = v, \\ v' = f(x, y, v), \end{cases} \qquad \begin{aligned} y(a) &= \alpha, \\ v(a) &= \gamma. \end{aligned}$$

This we can solve using a standard method for initial value problems, e.g., a Runge–Kutta method.

Since the given problem is a boundary value problem, we do *not* know $y'(a)$. Let us then guess a value of $y'(a)$, solve the initial value problem numerically, and compare the value of the numerical solution at $x = b$ with the given boundary value $y(b) = \beta$.

Example 10.11.1   Given the boundary value problem

$$y'' = -y, \qquad y(0) = 0, \qquad y(\pi/2) = 1$$

(which has the analytic solution $y(x) = \sin x$). From the formulation of the problem, we know that the solution passes through the two points in Figure 10.11.2, marked ×.

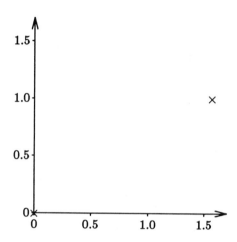

Figure 10.11.2   The boundary values marked ×.

The straight line between these points has the slope

$$\frac{1-0}{\pi/2 - 0} \approx 0.637.$$

Put $\gamma_0 = 0.637$. This should be a reasonable guess for $y'(0)$. Then, we have the initial value problem

$$y'' = -y, \qquad y(0) = 0, \qquad y'(0) = \gamma_0 = 0.637,$$

which we can write as a system

$$\begin{cases} y' = v, & y(0) = 0, \\ v' = -y, & v(0) = 0.637. \end{cases}$$

This we solve using Runge–Kutta's (classical) method and step length $h = \pi/100$ (i.e., we take 50 steps). The result is illustrated in Figure 10.11.3.

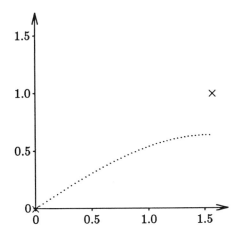

Figure 10.11.3   Solution of the initial value problem with $y'(0) = 0.637$.

We see that the computed solution is below the prescribed boundary value $y(\pi/2) = 1$. Therefore, we try a larger value of $y'(0)$, e.g., $y'(0) = \gamma_1 = 1.2$, and solve the corresponding initial value problem. The result is shown in Figure 10.11.4.

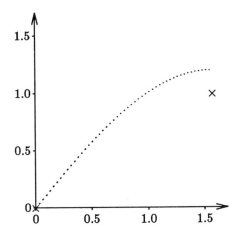

Figure 10.11.4   Solution of the initial value problem with $y'(0) = 1.2$.

In order to determine the value of the derivative at the left endpoint of the interval, we should be systematic. Given a guess for the derivative, $y'(a) = \gamma$, the initial value problem

$$y'' = f(x, y, y'), \qquad y(a) = \alpha, \qquad y'(a) = \gamma,$$

has a solution that depends on γ. We denote this solution $y(x, \gamma)$, and put

$$g(\gamma) = y(b, \gamma).$$

To determine the correct value of $y'(a)$ is equivalent to finding γ so that $y(b, \gamma) = \beta$. In other words: We will *solve the equation*

$$g(\gamma) - \beta = 0.$$

This is an equation where we cannot give an explicit expression for the function g, but we can compute approximations of function values $g(\gamma)$ by solving an initial value problem numerically. As we do not know the function g explicitly, we cannot differentiate it (*approximate* derivative values can be computed, however). To solve the equation $g(\gamma) - \beta = 0$, we should therefore use the *secant method*:

$$\gamma_{n+1} = \gamma_n - (g(\gamma_n) - \beta) \frac{\gamma_n - \gamma_{n-1}}{g(\gamma_n) - g(\gamma_{n-1})}.$$

Example 10.11.5   In the previous example, we have computed the following function values:

γ	$g(\gamma) - \beta$
0.637	−0.363
1.2	0.2

The secant method now gives

$$\gamma_2 = 1.2 - 0.2 \, \frac{(1.2 - 0.637)}{0.2 - (-0.363)} = 1.$$

With the initial value $y'(0) = \gamma_2 = 1$, we get $y(\pi/2, \gamma_2) = 1.0000$, which is the correct value.

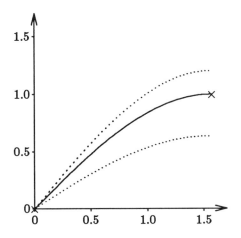

Figure 10.11.6 The solid curve is obtained with $y'(0) = 1$.

For obvious reasons, this method is called the **shooting method**.

It was no coincidence that already in the third try we had the correct value of $y'(0)$. One can show that if the function f in the right hand side of the differential equation is linear in y and y', then the function g is a first degree polynomial in γ. In the more general case, when f is nonlinear in y and y', then also g is nonlinear, and one must perform several iterations.

Example 10.11.7 The boundary value problem

$$y'' = 1 + yy', \qquad y(0) = 1, \qquad y(0.6) = 2,$$

is nonlinear in y and y', and therefore the function $g(\gamma)$ is nonlinear in γ. We use Runge–Kutta's (classical) method with the step length $h = 0.01$ to solve the corresponding initial value problems. We choose $\gamma = (2-1)/(0.6-0) \approx 1.67$ as a starting guess for $y'(0)$. With the secant method, we get

γ	$y(0.6, \gamma)$	$g(\gamma) - \beta$
1.67	2.879	0.879
0.8	1.955	−0.045
0.842	1.9957	−0.0043
0.846437	2.000113	0.000113

Exercises

1. Write the differential equation

$$3y''' + 4xy'' + \sin y = f(x)$$

   as a system of first order differential equations.

2. An *autonomous* system of differential equations is a system of the form $y' = f(y)$, where the independent variable (x or t, usually) does not enter explicitly into the right hand side. Introduce a new variable in a system $y' = f(x, y)$, and rewrite it in autonomous form.

3. Given the initial value problem

$$y' = y, \qquad y(0) = 1.$$

   a) Use Euler's method to compute an approximation of $y(x)$, i.e., take a step with Euler's method and use the step length $h = x$.
   b) Do the same with Heun's and Runge–Kutta's methods. Compare to the Maclaurin expansion of the solution y.

4. Use Euler's method and Richardson extrapolation to compute an approximation of $y(0.2)$, where $y(x)$ is the solution of the initial value problem $y' = x^2 - y$, $y(0) = 2$. Use the step lengths 0.2 and 0.1.

5. Solve the following two initial value problems numerically, using the trapezoidal method.
   a) $y' = 2x - y$, $y(0) = 1$. Use $h = 0.2$, and compute an approximation of $y(1)$.
   b) $y' = xy^2$, $y(0) = 1$. Use $h = 0.5$, and compute an approximation of $y(0.5)$.

6. Given the initial value problem

$$y'' = 2y(1 + y^2), \qquad y(0) = 1, \qquad y'(0) = 2.$$

   a) Write the equation as a system of first order equations.
   b) Compute an approximation of $y(0.4)$ using Euler's method. Use $h = 0.2$ and $h = 0.1$, and perform Richardson extrapolation.

7. The initial value problem

$$\begin{cases} x' = x(1 - y), & x(0) = 5, \\ y' = y(0.75x - 1.5), & y(0) = 2, \end{cases}$$

is a predator–prey model, where $x(t)$ is the number (in some unit, say thousands) of individuals of a prey animal and $y(t)$ is the number of individuals of a predator animal at time t. Integrate the system to $t = 1$ using Euler's method and $h = 0.1$.

8. Given the difference equation

$$y_{n+3} + by_{n+2} + cy_{n+1} + dy_n = 0.$$

Assume that the corresponding characteristic equation has a double root r_1 and a simple root r_2. Show that the general solution is

$$y_n = Anr_1^{n-1} + Br_1^n + Cr_2^n.$$

Give the general solution for the case when the characteristic equation has a triple root.

9. Write a computer program for Runge–Kutta's method, and solve numerically the test problem

$$y' = \lambda y, \qquad y(0) = 1.$$

Put $\lambda = -1000$, and determine the maximal step length h_{max}, for which the method gives a decreasing solution. Derive the difference equation for the method applied to the test equation. Put $\lambda = -1000$ and $h = h_{max}$ in the characteristic polynomial, and verify that it has a root approximately equal to 1.

10. Given the boundary value problem

$$y'' + p(x)y' - q(x)y = 0, \qquad y(a) = \alpha, \qquad y(b) = \beta,$$

where $q(x) \geq 0$, and $|p(x)| \leq P$ for $x \in [a, b]$. Discretize the derivatives using central differences, and write down the corresponding system of equations. Show that the matrix is diagonally dominant if $h \leq 2/P$.

11. Solve the boundary value problem

$$y'' - x^2 y = 0, \qquad y(0) = 0, \qquad y(1) = 1,$$

using the difference method with $h = 0.25$.

12. Solve the boundary value problem

$$y'' = x - 2y, \qquad y(0.4) = 0.3, \qquad y(0.8) = 0.7,$$

using the shooting method, and a suitable method and step length for the initial value problem.

13. Solve the boundary value problem

$$-y'' + 2y = 2, \qquad y(0) = y(1) = 0,$$

using Galerkin's method with linear elements and $h = 0.25$.

14. What is the stiffness matrix for Galerkin's method applied to the boundary value problem

$$-y'' + qy = f, \qquad y(0) = y(6) = 0,$$

using linear elements over the net $(0, 1, 2, 4, 6)$?

References

Numerical methods for the solution of ordinary differential equations have a long history. Euler made basic contributions around 1760. The classical Runge–Kutta method was invented at the end of the nineteenth century. During the 1960s and 1970s, efficient methods for stiff systems were found. A good introduction to classical and modern methods is given in

C. W. Gear, *Numerical Initial Value Problems in Ordinary Differential Equations*, Prentice–Hall, Englewood Cliffs, New Jersey, 1971.

Boundary value problems are treated extensively in

H. B. Keller, *Numerical Methods for Two-Point Boundary Value Problems*, Blaisdell Publishing Comp., Waltham, Massachusetts, 1968.

As finite element methods have numerous important applications, there is a vast literature in the area. The following two books are mathematically–numerically oriented, and mostly treat the solution of partial differential equations.

C. Johnson, *Numerical Solution of Partial Differential Equations by the Finite Element Method*, Studentlitteratur, Lund, 1987.

G. Strang and G. J. Fix, *An Analysis of the Finite Element Method*, Prentice-Hall, Englewood Cliffs, New Jersey, 1973.

A somewhat more application-oriented presentation is given in

J. N. Reddy, *An Introduction to the Finite Element Method*, McGraw-Hill, New York, 1984.

Answers to Exercises

Chapter 1

3. Boundary conditions: $y(0) = y''(0) = y(1) = y''(1) = 0$;
   $y(x) = q_0/(24EI_0)(x^4 - 2x^3 + x)$.

Chapter 2

2. Absolute error less than $1.7 \cdot 10^{-4}$.

4.
$$\Delta y \approx \Delta x/x, \qquad \frac{\Delta f}{f} \approx \alpha_1 \frac{\Delta x_1}{x_1} + \alpha_2 \frac{\Delta x_2}{x_2} + \alpha_3 \frac{\Delta x_3}{x_3}.$$

5. $f = 18.9 \pm 0.6$ mm.

6. $h = 0.049 \pm 0.0041$ km.

8. a) $e^x - e^{-x} = 2(x + x^3/3! + x^5/5! + \ldots)$.
   b) $\sin x - \cos x = -\cos 2x/(\sin x + \cos x)$.
   c) $1 - \cos x = 2\sin^2 \frac{x}{2}$.
   d) $(\sqrt{1 + x^2} - \sqrt{1 - x^2})^{-1} = (\sqrt{1 + x^2} + \sqrt{1 - x^2})/(2x^2)$.

11.
   if $|x_1| > |x_2|$ **then**
   $\qquad t := x_2/x_1; \text{ sq} := |x_1|\sqrt{1 + t^2}$
   else
   $\qquad t := x_1/x_2; \text{ sq} := |x_2|\sqrt{1 + t^2}$
   endif

14.   Forward: $|\hat{S}_n - S_n| \leq \sum_1^n |x_i \delta_i|$, $|\delta_i| \leq 1.06k\mu$.
       Backward: $\hat{S}_n = \sum_1^n x_i(1 + \delta_i)$, $|\delta_i| \leq 1.06k\mu$.

Chapter 3

1. a) $\sin 0.1 = 0.100 \pm 0.2 \cdot 10^{-3}$.
   b) $e^{0.12} = 1.1272 \pm 0.3 \cdot 10^{-3}$ (compare with geometric series).
2. $S = 1.2026 \pm 31 \cdot 10^{-4}$.
3. $S = 0.514 \pm 0.3 \cdot 10^{-3}$.
4. $|\Delta u|/|u| \lesssim (1/|u|)((x_0 - n\pi_0) + 3nr)\mu \leq 2.3\mu \leq 1.4 \cdot 10^{-7}$.
5. For $|x_0 - \pi| \leq 1.76 \cdot 10^{-8}$.
7. $\tau = 0.60725293500888$.
8. For $2^{-3i}/3 < 2^{-46}$, which gives $i \geq 15$.

Chapter 4

3. $x^* = 0.426302 \pm 0.8 \cdot 10^{-6}$.
4. The attainable accuracy is $5.6 \cdot 10^{-8}$.
5. $x^* \approx 0.42$ and the attainable accuracy is $5.9 \cdot 10^{-6}$. $x^* = 0.4172373 \pm 60 \cdot 10^{-7}$.
6. $\lambda = 0.87$ and $x^* = 0.510973 \pm 0.5 \cdot 10^{-6}$.
8. $x^* = 0.523596 \pm 0.3 \cdot 10^{-5}$.
10. $f(x) = (x - 8.25)(16x^2 - 12)$.
11. $|x_9 - x^*| \leq 0.0032$.
12. Differentiation gives $(y + xy')\cos(xy) - y' + 1 = 0$. Since the maximum occurs for $y' = 0$, we get the second equation in the system. $(x_1, y_1) = (1.0797, 1.9454)$.

Chapter 5

1. a) The table size is 3073, $h = 2^{-10} \leq \sqrt{2} \cdot 2^{-10}$.   b) Table size 1536.
2. a) 684.895.   b) 666.666.   c) Error in sin and cos is $\leq 0.5 \cdot 10^{-6}$. Error in cot ≤ 0.24.   d) Interpolation close to the singularity at the origin for cot.

3. Use the fact that the difference is the coefficient of x^k in the interpolating polynomial of degree k to f through (x_i, f_i), $i = 1, 2, \ldots, k + 1$.

4. $16 \cdot 10^{-d}$.

5.
$$\int_a^b f(x)\,dx \approx \frac{b-a}{n-1}\left(\frac{1}{2}f_1 + f_2 + \ldots + f_{n-1} + \frac{1}{2}f_n\right).$$

7. $f(2.5) \approx 3.725$.

8. a)
$$s(x) = \begin{cases} 1 - 1.4x + 0.4x^2, & 0 \leq x \leq 1, \\ -0.2(x-1) + 1.2(x-1)^2 - 0.5(x-1)^3, & 1 \leq x \leq 2, \\ 0.5 + 0.7(x-2) - 0.3(x-2)^2 + 0.1(x-2)^3, & 2 \leq x \leq 3. \end{cases}$$

b) $s(0.5) = 0.35$, $s(1.5) = 0.1375$, $s(2.5) = 0.7875$.

c) $s'(3) = 0.4$.

Chapter 6

1. $h_{\text{opt}} = 2.8 \cdot 10^{-3}$. Total error $\leq 1.2 \cdot 10^{-5}$.

3. $f'(1) \approx 0.540300$ (correct: 0.540302).
   $f''(1) \approx -0.841511$ (correct: -0.841471).

4. The truncation error in $D(0.05)$ can be estimated as

$$R_T \lesssim |D(0.05) - D(0.1)| = |0.1224267 - 0.1224217| \leq 5 \cdot 10^{-6}.$$

It does not pay off to use Richardson extrapolation because of the magnitude of R_{XF}.

5. The acceleration is 2.31 ± 0.18 m/s^2.

Chapter 7

1. $T(0.5) + \frac{1}{3}(T(0.5) - T(1)) = S(0.5)$, and $I = S(h) - Ch^4 f^{(4)}(\eta)$.
   Here, $f(t) = t^3$ and $I = S(h)$.

2. 1.423 ± 10^{-2} liters.

3. a) $F(1) = 0.746825 \pm 11 \cdot 10^{-6}$.
   $|R_T| \lesssim 8 \cdot 10^{-6}$, $|R_{\text{XF}}| \lesssim 0.5 \cdot 10^{-6}$, $|R_B| \lesssim 2 \cdot 10^{-6}$.
   b) 0.7ϵ.

4. $I = 0.333 \pm 5 \cdot 10^{-3}$. Subintervals: $(0,1)\,(1,2)$, and $(2,4)$.

5. a) $R_T = a_1 h^{3/2} + a_2 h^2 + a_3 h^{5/2} + \dots$.

   b) If we use a), we get the approximation 0.188015. The correct value rounded to eight decimals is 0.18801536.

6. The substitution $x = \sin t$ gives $I = \int_0^{\pi/2} \sin(\sin t)\,dt$, which can be computed by Romberg's method. Alternatively, series expansion gives

$$I = \sum_{n=0}^{\infty} (-1)^n \frac{1}{((2n+1)!!)^2}.$$

7. The substitution $t = e^{-x}$ gives $I = \int_0^1 (1 - t \log t)^{-1}\,dt$.

Chapter 8

1. $(x_1, x_2, x_3) = (1.2, -0.4, 0.4)$.

2.

$$P = \begin{pmatrix} 0 & 0 & 0 & 1 & 0 \\ 1 & 0 & 0 & 0 & 0 \\ 0 & 0 & 1 & 0 & 0 \\ 0 & 1 & 0 & 0 & 0 \\ 0 & 0 & 0 & 0 & 1 \end{pmatrix}.$$

3. a) $x = (1, 2, 3)^T$.

   b)

$$P = \begin{pmatrix} 0 & 0 & 1 \\ 1 & 0 & 0 \\ 0 & 1 & 0 \end{pmatrix}, \quad L = \begin{pmatrix} 1 & 0 & 0 \\ 0.4 & 1 & 0 \\ 0.3 & 0.6 & 1 \end{pmatrix}, \quad U = \begin{pmatrix} 2 & 1 & 0 \\ 0 & 1 & 3 \\ 0 & 0 & 1 \end{pmatrix}.$$

4. a)

$$P = \begin{pmatrix} 0 & 1 & 0 \\ 0 & 0 & 1 \\ 1 & 0 & 0 \end{pmatrix}, \quad L = \begin{pmatrix} 1 & 0 & 0 \\ 0.2 & 1 & 0 \\ 0.7 & 0.6 & 1 \end{pmatrix}, \quad U = \begin{pmatrix} 2 & 1 & 1 \\ 0 & 1.2 & 3 \\ 0 & 0 & 4 \end{pmatrix}.$$

   b) $Ly = Pb$ gives $y = (11, 19.8, 20)^T$, and $Ux = y$ gives $x = (1, 4, 5)^T$.

5.

$$L = \begin{pmatrix} I & 0 \\ p & 1 \end{pmatrix}, \quad U = \begin{pmatrix} D & q \\ 0 & 8.6 \end{pmatrix},$$

where $p = (0.5, -0.2, 0.5, -0.2)$, $q = (1, -1, 1, -1)^T$, and D is the diagonal matrix $D = \text{diag}(2, 5, 2, 5)$. Solution: $x = (4.5, 1.2, 2.3, 3.4, 0.1)^T$.

6. a)
$$L = \begin{pmatrix} 1 & 0 & 0 & 0 \\ 0.5 & 1 & 0 & 0 \\ 0 & 0.5 & 1 & 0 \\ 0 & 0 & 0.5 & 1 \end{pmatrix}, \quad D = \mathrm{diag}(4,4,4,4).$$

b)
$$U = \begin{pmatrix} 2 & 1 & 0 & 0 \\ 0 & 2 & 1 & 0 \\ 0 & 0 & 2 & 1 \\ 0 & 0 & 0 & 2 \end{pmatrix}$$

7. $x = (0.6523, -0.1080, 0.3793)^T$.

8. $p(x) = 2.3327 - 0.2230x + 0.0061x^2$.

12. $\kappa_\infty(A) = 144, \ \|\delta x\|_\infty/\|x\|_\infty \leq 0.072$.

13. a)
$$P = \begin{pmatrix} 0 & 1 & 0 \\ 0 & 0 & 1 \\ 1 & 0 & 0 \end{pmatrix}, \ L = \begin{pmatrix} 1 & 0 & 0 \\ -1 & 1 & 0 \\ 10^{-3} & 0.4995 & 1 \end{pmatrix}, \ U = \begin{pmatrix} 1 & 1 & 1 \\ 0 & 2 & 2 \\ 0 & 0 & -2 \end{pmatrix}.$$

b)
$$A^{-1} = \begin{pmatrix} 0 & 0.5 & -0.5 \\ 0.5 & 0.25 & 0.25 \\ -0.5 & 0.25 & 0.25 \end{pmatrix}.$$

(Rounded to 2 decimals.)

c) $\kappa_\infty(A) = 3, \ \|\delta x\|_\infty/\|x\|_\infty \leq 0.0027$.

14. a)
$$L = \begin{pmatrix} 1 & 0 & 0 \\ 0.15 & 1 & 0 \\ 0.20 & 0.11 & 1 \end{pmatrix}, \quad U = \begin{pmatrix} 20 & 3.0 & 4.0 \\ 0 & 40 & 4.4 \\ 0 & 0 & 59 \end{pmatrix}.$$

b) $r^{(1)} = (0.20, -9.0, -1.6)^T, \ x^{(2)} = (0.74, -10, 7.8)^T$.

Chapter 9

6. a) $a_{n+1} = 2, b_{n+1} = 0, c_{n+1} = 1, n \geq 1.$ $U_0(x) = 1, U_1(x) = 2x.$

c) $w(x) = \sqrt{1 - x^2}$.

7. Choose P_{n-1} equal to the polynomial that interpolates f at the zeros of $T_n(x)$.

9. a) $\frac{3}{14}(2 + 3x)$, the norm of the error function: 0.055.

b) $1/(3\sqrt{3}) + x$, the norm of the error function: $1/(3\sqrt{3})$.

10. a) $\frac{3}{35}(40x^2 - 40x + 9)$, the norm of the error function: $8/105$.

b) $4x^2 - 4x + 7/8$, the norm of the error function: $1/8$.

11. a) $\varphi_0 = 1$, $\varphi_1 = x$, $\varphi_2 = x^2 - 3/5$.

b) $p_2(x) = \sum_{k=0}^{2} c_k \varphi_k(x)$,

$c_0 = \frac{3}{2}(e - 5e^{-1})$, $c_1 = 5(-e + 8e^{-1})$, $c_2 = \frac{175}{4}(\frac{21}{5}e - 31e^{-1})$.

c) $\frac{1}{4}(e^2 - 5e^{-2}) - (\frac{2}{3}c_0^2 + \frac{2}{5}c_1^2 + \frac{8}{175}c_2^2) \approx 0.00058$.

12. a) $\varphi_0 = 1$, $\varphi_1 = x$, $\varphi_2 = x^2 - 2$, $\varphi_3 = x^3 - 3.4x$, $\varphi_4 = x^4 - 31x^2/7 + 72/35$.

b) No. The polynomial $\varphi_5 = c(x+2)(x+1)x(x-1)(x-2)$ is orthogonal to $\varphi_0, \ldots, \varphi_4$, but it has $\|\varphi_5\| = 0$.

c) $p^*(x) = x^4 - x^3 + 5$.

Chapter 10

3. a) Euler: $1 + x$.

b) Heun: $1 + x + x^2/2$, and Runge-Kutta: $1 + x + x^2/2! + x^3/3! + x^4/4!$.

4. $y(0.2; h = 0.2) = 1.6$, $y(0.2; h = 0.1) = 1.621$. $y(0.2)_{\text{extr}} = 1.642$.

5. a) $y(1) \approx 1.09995$.

b) $y(0.5) \approx 1.17157$.

6. a) With $z = y'$, we get

$$\begin{cases} y' = z, & y(0) = 1, \\ z' = 2y(1 + y^2), & z(0) = 2. \end{cases}$$

b) $y(0.4; h = 0.2) = 1.96$, and $y(0.4, h = 0.1) = 2.126$. $y(0.4)_{\text{extr}} = 2.292$.

7. $x(1) \approx 0.460$, $y(1) \approx 2.51$.

11. $y \approx (0.2392, 0.4793, 0.7269)$.

12. With Euler's method and the step length $h = 0.2$, we obtain: $\gamma_0 = 1.0$ gives $y(0.8; \gamma_0) = 0.692$. The next guess, $\gamma_1 = 1.1$, gives $y(0.8; \gamma_1) = 0.732$. Using the secant method, we then get $\gamma_2 = 1.02$, which gives $y(0.8; \gamma_2) = 0.7$.

13. $y(0.25) = y(0.75) \approx 0.1580$, $y(0.5) \approx 0.2085$.

14.

$$K_1 = \begin{pmatrix} 2 & -1 & 0 \\ -1 & 1.5 & -0.5 \\ 0 & -0.5 & 1 \end{pmatrix}.$$

Index

A

absolute error, 9
Adams–Bashforth, 305
Adams–Moulton, 305
algebraic equation,89
algorithm, 5
alternating series, 50
alternation property, 281
approximation problem, 253
asymptotic error constant, 84
attainable accuracy, 87

B

back substitution, 175
backward analysis, 31, 223
backward difference, 142
band matrix, 208
band matrix method, 317
base (for number system), 19

basic (floating point) format, 33
beam equation, 3
bisection method, 73
boundary condition, 3, 314
boundary value problem, 3, 314
boundary values, 314
B-spline, 130

C

cache memory, 242
cancellation, 18
central difference, 143
chaining, 42
characteristic equation, 305
Chebyshev
 approximation, 275
 interpolation, 275
 nodes, 272
 norm, 250, 253